Epigenetics in Crop Improvement

Luis María Vaschetto

Epigenetics in Crop Improvement

Safeguarding Food Security in an Ever-Changing Climate

 Springer

Luis María Vaschetto
The Ronin Institute for Independent Scholarship
Montclair, NJ, USA

ISBN 978-3-031-73178-5   ISBN 978-3-031-73176-1   (eBook)
https://doi.org/10.1007/978-3-031-73176-1

This Springer imprint is published by the registered company Springer Nature Switzerland AG
The registered company address is: Gewerbestrasse 11, 6330 Cham, Switzerland

If disposing of this product, please recycle the paper.

Preface

Plants are complex biological systems with an intrinsic resilience to face adverse environmental conditions. Over millions of years, plants have developed the ability to promptly tackle environmental challenges. They do this not only by depending on the selection of adaptive gene variants but also by employing alternative mechanisms even when genetic variation is lacking. Over the last decades, advancements in plant biology have elucidated outstanding molecular mechanisms underlying environmentally driven plasticity in complex plant traits, evidencing how we can exploit variation beyond genetics to ensure food security for the coming generations.

All multicellular organisms including plants rapidly respond to external stimuli though epigenetic pathways, that is, via heritable changes in gene expression that do not involve changes in the DNA sequence. Epigenetic information systems are encrypted as chemical marks on DNA and associated histones, as well as through the transgenerational transmission of non-coding RNAs (ncRNAs), ultimately shaping gene expression patterns and phenotypes. This information can even be transmitted to offspring, challenging traditional Mendelian genetics and bringing up fascinating biological phenomena such as paramutation, genomic imprinting, and transgenerational epigenetic inheritance.

Epigenetics in Crop Improvement: Safeguarding Food Security in an Ever-Changing Climate explores the amazing potential of epigenetics in developing more resilient crops and addresses the challenges faced by a changing environment. This book delves into the dynamic patterns of non-canonical variation, revealing how epigenetics mechanisms shape plant growth, their development, and adaptation, and examining the potential for developing crops better adapted to ever changing climate conditions. To guide readers through this scientific frontier, the book also includes a comprehensive glossary of key epigenetic terms.

Montclair, NJ, USA Luis María Vaschetto

Contents

Part I
Introduction

Chapter 1
Epigenetics and its Role in Plant Evolution

Epigenetics is defined as a group of chemical modifications that occurs independently of alterations in the underlying DNA sequence and has the potential to alter gene expression. This field encompasses rearrangements within the DNA molecule, impacting its structure, organization, and associated proteins, ultimately leading to changes in gene activity. In plants, epigenetics stands as a foundational pillar of adaptability and resilience, which is achieved through the meticulous storage of information via information systems that translate into chemical modifications on DNA and histones, as well as the action of non-coding RNAs (Fig. 1.1). These epigenetic information systems operate in concert to regulate gene expression patterns, guiding cellular differentiation, controlling development, and fine-tuning responses to environmental signals (Vaschetto, 2015).

DNA methylation, a covalent modification where a methyl group (CH_3) is generally added to the cytosine base in DNA through the activity of specific enzymes called DNA methyltransferases, has the potential to silence or activate gene expression by allowing or inhibiting factor transcription binding. Similarly, histone modifications, which include among others acetylation, methylation, phosphorylation, and ubiquitination, can also exert regulatory effects on gene expression either by hindering the binding of transcription factors or by recruiting ribonucleoprotein complexes that modify the structure of chromatin, the complex of DNA and proteins that package DNA into chromosomes. These histone marks can profoundly influence the accessibility of DNA, ultimately modifying gene activity. Non-coding RNAs (ncRNAs), including small ncRNAs (sncRNAs) and long non coding RNAs (lncRNAs) also play a crucial role in epigenetics. By recruiting specific proteins that modify chromatin, ncRNAs can switch genes on or off without altering the DNA sequence itself. This epigenetic control allows cells to fine-tune gene expression, creating diverse cell types in higher organisms and responding to environmental cues.

L. M. Vaschetto, *Epigenetics in Crop Improvement*, https://doi.org/10.1007/978-3-031-73176-1_1

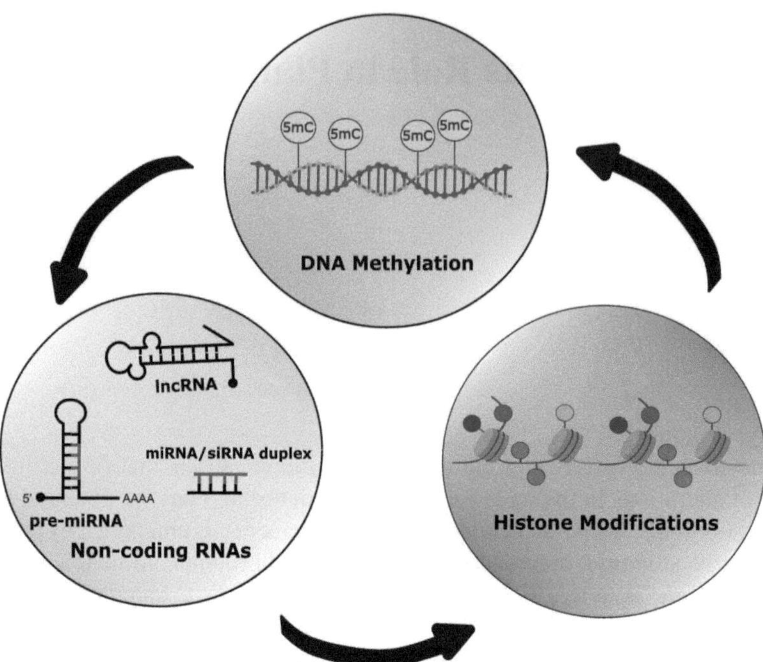

Fig. 1.1 Epigenetic Information Systems. The model depicts the three epigenetic information systems. DNA methylation involves the addition of methyl groups to specific DNA cytosine bases. Histone modifications, i.e., chemical modifications to histone proteins that can alter chromatin structure and accessibility, encompass a diverse range of epigenetic modifications, including acetylation, methylation, and ubiquitylation, each influencing chromatin structure and gene expression. Finally, regulatory non-coding RNAs such as small interfering RNAs (siRNAs) and long non-coding RNAs (lncRNAs) are capable of modulating gene expression patterns through various mechanisms, including chromatin remodeling and transcriptional silencing. The three epigenetic systems are interdependent, meaning that the establishment of one epigenetic mark can trigger the action of the other two systems. This creates a feedback loop that reinforces the overall epigenetic state. For instance, the establishment of repressive DNA methylation marks on the promoter region of a target gene by a non-coding RNA-guided complex can induce the recruitment of histone deacetylases. These enzymes remove acetyl groups from histones, leading to a more condensed chromatin structure and reduced gene expression. Consequently, the interplay between epigenetic systems further reinforce the repressive state through heterochromatin formation

Natural Selection, Evolution, Phenotype, and Epigenetics

Natural selection, the primary driving force behind evolution, favors organisms with traits that enhance their survival and reproductive success. Epigenetic modifications can exert significant influence on the expression of genes responsible for environmental adaptation, potentially granting a competitive advantage to individuals harboring beneficial epigenetic variants (Douhovnikoff & Dodd, 2015). As our understanding of epigenetics continues to evolve, a clearer picture emerges of how these modifications shape the course of evolution and enrich our comprehension of this dynamic process.

Phenotypic plasticity allows a single genotype to generate a diverse array of phenotypes tailored to specific environmental conditions. This characteristic serves as a critical adaptive strategy for plants to maximize fitness and survival, despite possessing a singular genotype. Remarkably, epigenetic marks can be transmitted mitotically, between mother and daughter cells, and meiotically, between parents and offspring. Meiotic epigenetic inheritance involves transgenerational inheritance, a phenomenon through which epigenetic modifications are passed down from an organism to its offspring, and potentially even further down the generational line. These types of modifications hold the potential for rapid adaptation to changing environments, facilitating evolutionary responses.

Beyond its impact on phenotypic plasticity, emerging evidence also highlights the role of epigenetic modifications in speciation, the process through which new species arise. By influencing gene expression and leading them down divergent paths within a population, epigenetic modifications can contribute to the establishment of reproductive isolation (Gehring, 2019). This interplay between genetic and epigenetic factors can further drive the development of distinct morphological, physiological, and behavioral characteristics. Ultimately, these disparities can render interbreeding between individuals from different populations increasingly challenging, contributing to the emergence of new species.

Inheritance of Epigenetic Modifications

As mentioned above, epigenetic modifications may be associated with two distinct processes: mitotic inheritance and meiotic inheritance. Through mitotic divisions, epigenetic marks are faithfully replicated and transmitted to daughter cells, ensuring their stable maintenance throughout the life cycle of a multicellular organism. In contrast, meiosis, a crucial process for sexual reproduction in which a single cell divides twice to form germinal cells, presents an opportunity for epigenetic reprogramming. Plant somatic cells undergo extensive epigenetic reprogramming during the transition to reproductive fate, giving rise to pluripotent gametophytes (She et al., 2013). This cellular process, characterized by global erasure and subsequent reestablishment of epigenetic marks within germline cells, offers a mechanism for generating epigenetic diversity (Feng et al., 2010). The reprogramming process is a major barrier for transgenerational epigenetic inheritance, where epigenetic modifications persist during meiosis to be inherited in the next generation, ultimately contributing to the potential for environmentally acquired characteristics to be passed on to future generations. It is important to highlight that plants may exhibit limited DNA methylation reprogramming across generations, leading to the inheritance of aberrant methylation patterns due to the failure to re-establish this mark during sexual reproduction (Quadrana & Colot, 2016).

Evolutionary Epigenetics

While transgenerational epigenetic inheritance deviates from the classical considerations of Lamarckian evolution, it nevertheless offers a plausible mechanism whereby environmentally mediated phenotypes can be partially selected to be transmitted across generations (Paszkowski & Grossniklaus, 2011). This raises the viewpoint of transgenerational epigenetic inheritance contributing to the intricate dynamics of evolution. As a result, epigenetic inheritance is currently a hot topic in evolutionary biology. In crop improvement, similar to other fields, epigenetic marks hold immense potential. By strategically manipulating and inducing these marks, scientists can engineer crops with desirable and adaptable traits.

Cells possess a remarkable capacity to maintain and pass on their distinct patterns of gene activity to their progeny. This "epigenetic memory"is intrinsically influenced by environmental stimuli and the underlying genetic architecture (Kinoshita & Seki, 2014). Certain epigenetic modifications exhibit remarkable resilience, defying reprogramming during developmental transitions (Tao et al., 2017). These epigenetic marks may potentially contribute to transgenerational phenotypic stability and evolutionary adaptation.

The dynamic nature of epigenetic modifications underscores the need to differentiate between those exhibiting varying degrees of plasticity and stability. Certain modifications, characterized by their responsiveness to environmental cues or developmental signals, demonstrate exceptional adaptability (Rehman & Tanti, 2020). These "plastic" modifications, capable of sustaining reversible changes and temporal variations, allow plants to respond rapidly to changing environmental conditions (Rajpal et al., 2022). The intricate and dynamic interplay between plastic and stable modifications may potentially be the basis for short-term adaptability and phenotypic agility, enabling plants to thrive in a constantly evolving environment.

Epigenetic Variation

Epigenetic mechanisms play a central role in mediating organism responses to environmental stimuli and facilitate adaptation to changing conditions. Environmental conditions can trigger 'epigenetic switches', generally in concert with genetic-mediated pathways, and it leads to adaptive phenotypic plasticity (Pimpinelli & Piacentini, 2020). The resulting epigenetic modifications can in turn exhibit exceptional sensitivity to specific environmental inputs (e.g., extreme temperatures) and thus exert regulatory influence over stress responses, developmental processes, and seasonal phenotypes (Thiebaut et al., 2019; Chang et al., 2020).

Epigenetic variation contributes to the success of species, particularly under stress conditions (Mounger et al., 2021), and it has been associated with evolutionary mechanisms like hybridization and genome duplication (Paun et al., 2007). Intriguingly, epigenetic variation exists in the inducibility, persistence, and stability of transgenerational environmental effects (Alvarez et al., 2020), suggesting their potential as independent traits. Consequently, genetics and epigenetics operate on distinct yet complementary pathways, working together to guide development programs. Later in this book, we will explore how epigenetic and genetic pathways interact, with a focus on the role of key transcription factors that integrate signals from both internal and external environments. Ultimately, epigenetic variation would serve as a complementary tool to cope with environmental changes, contributing significantly to their stability and survival.

> **Box 1.1. Epigenetics and Genetics: It Takes Two to Tango**
> For a comprehensive understanding of how gene expression mechanisms operate at molecular level, it is crucial to distinguish classical genetic from epigenetic pathways. While they are complementary, epigenetic information systems exert its influence through mechanisms that do not involve direct changes in DNA sequence. In the intricate epigenetic pathways, specific nucleotide sequence contexts interact with the epigenetic machinery upon both endogenous and exogenous environmental factors to modulate gene expression. For instance, tandem repeat sequences, short nucleotide copies which lie adjacent to each other and involve a repetitive unit, have shown to be susceptible to the epigenetic phenomenon of paramutation through which alleles *trans*-interact to trigger mitotically and meiotically heritable alterations in gene expression. The concept 'epigenetic-sequence context' highlights the critical role of non-coding regulatory DNA regions in determining epigenetic mechanisms and the subsequent heritability of epigenetic modifications. Ultimately, specific nucleotide sequence contexts influence the tridimensional (3D) genomic architecture, thereby also impacting the potential for gene regulation via canonical genetic pathways (e.g., transcription factor binding).

Within the dynamic interplay of genetic and environmental factors, epigenetic modifications emerge as critical regulators of plant stress tolerance. DNA methylation and histone modifications orchestrate gene expression patterns in response to environmental stressors (e.g., drought). For instance, alterations in the DNA methylation profile can modulate the expression of stress-responsive genes (Garg et al., 2015; Wang et al., 2021), allowing plant survival under adverse conditions. Nevertheless, the significance of epigenetic regulation extends far beyond its role in mediating stress responses. DNA methylation and histone modifications also act as dynamic regulators, meticulously fine-tuning various developmental processes, such as the precise timing of floral initiation, in response to environmental cues like temperature and photoperiod (Burn et al., 1993; Shi et al., 2023). These dynamic epigenetic modifications ensure that plants adapt their life cycles with remarkable precision to match an ever-changing environmental landscape, ultimately maximizing reproductive success.

Epigenetic Memory

Beyond the immediate impact on the individual, epigenetic changes exert a profound influence across generations through the formation of transgenerational stress memory. When plants encounter stress, they acquire modified epigenetic states, which may potentially be transmitted to subsequent generations (Thiebaut et al., 2019). This transgenerational stress memory provides the progeny with a crucial advantage, conferring an enhanced stress tolerance and allowing them to thrive under similar challenging conditions, showcasing the remarkable adaptive potential of nature. Epigenetic inheritance exhibits varying degrees of transmissibility and it may be associated with the type of sequence linked to the target epigenetic pathway. This is a consequence of evolution and the evolutionary forces that have shaped epigenetic variants.

Epigenetic Modifications Acting During Development

Epigenetic modifications hold a pivotal role in controlling cell fate and differentiation, acting as master conductors throughout plant development. They meticulously control gene expression patterns, assigning unique identities to cells and ensuring the establishment of a robust lineage memory. This regulatory developmental process is achieved through several mechanisms:

- Chromatin accessibility: Epigenetic modifications (e.g., histone modifications) act as molecular switches that control chromatin accessibility. The epigenetic landscape precisely regulates the access of transcription factors and regulatory proteins to genes, profoundly impacting cellular decisions and guiding differentiation. Chromatin-based mechanisms play a pivotal role in plant adaptation by tightly regulating cell fate acquisition and maintenance, both within and beyond meristematic regions. Specific epigenetic factors act as sophisticated regulatory networks, orchestrating stem cell activity, counting cell divisions, and positioning cell fate transitions through their sensitivity to phytohormone gradients (Morao et al., 2016).
- Flowering plants (angiosperms) utilize a sophisticated epigenetic reprogramming program during reproductive development. This dynamic process involves the precise erasure and reinforcement of epigenetic marks, ultimately shaping distinct gene expression patterns. These patterns dictate cell fate and ensure the faithful execution of differentiation trajectories, leading to the formation of the two highly similar male gametes and the two functionally distinct female gametes (Gutierrez-Marcos & Dickinson, 2012).
- Maintenance of cell identity: Once committed to specific lineages, cells rely on epigenetic modifications to preserve and stabilize their identities. This is achieved through the consolidation of established gene expression patterns. The presence of epigenetic marks ensures faithful inheritance of these patterns during mitosis,

maintaining lineage memory across generations. However, the remarkable plasticity of plant cells remains an intriguing question. While chromatin dynamics in the meristem influence gene expression, a definitive epigenetic constraint on cell fate appears to be established only during later stages of differentiation (Birnbaum & Roudier, 2017). This raises the possibility that plant cell plasticity stems from either the ability to erase pre-existing fate identity or the existence of specialized cells lacking a strong epigenetic barrier to fate reprogramming (Birnbaum & Roudier, 2017).

Adaptative Epigenetics: Vernalization as Study Case

A clear example of environmentally driven responses via epigenetic pathways is vernalization, the process by which plants are induced to flower in response to cold temperatures. Upon cold exposure, specific epigenetic modifications occur on the genes responsible for flowering, allowing the plant to "remember" cold conditions and consequently initiate flowering once the appropriate temperature and day length conditions are met. This memory, however, is not transmitted to the next generation, suggesting a resetting mechanism. The biological purpose of vernalization is to promote flowering during more favorable environmental conditions, maximizing reproductive success. In the model plant *Arabidopsis thaliana, FLOWERING LOCUS C* (*FLC*), a MADS box transcription factor that acts as a floral repressor gene, plays a crucial role in regulating flowering time (Michaels & Amasino, 1999; Henderson-Carter et al., 2023). This gene is a major target for critical epigenetic mechanisms. In this plant species, vernalization leads to the epigenetic silencing of *FLC*, allowing the expression of *FLOWERING LOCUS T* (*FT*), a flowering activator, ultimately triggering the transition to flowering (Finnegan et al., 2021). Chapter 3 analyzes in more detail the intricate epigenetic underpinnings of vernalization.

Epigenetic Diversity and Enhanced Traits

Two plants, despite sharing identical DNA sequences, can exhibit distinct phenotypic variations due to differences in their epigenetic landscapes, ultimately impacting on gene expression. As a result, epigenetic diversity represents a valuable reservoir of variation, allowing for fine-tuning of phenotypic characteristics in response to specific environmental conditions. This diversity manifests as "epialleles," distinct epigenetic modifications at specific genetic *loci*. Epimutations can be inherited from parents or arise entirely *de novo*. Analogous to bookmarks placed throughout a book, epimutations mark specific genes, influencing their expression levels through mechanisms independent of DNA sequence. The resulting epialleles often play crucial roles in shaping complex traits, including flower morphology, height, and environmental stress resistance (Jacobsen & Meyerowitz, 1997;

Jacobsen et al., 2000; Miura et al., 2009; Zhang et al., 2012; Stokes et al., 2002). Harnessing the power of epialleles holds immense potential for developing crops with enhanced resistance to pests, diseases, and the challenges posed by climate change (Gupta & Salgotra, 2022).

Epigenetic modifications also hold potential for improving agricultural resilience. Studies have demonstrated their effectiveness in enhancing stress tolerance in crop plants, as exemplified by the manipulation of stress-responsive genes leading to drought tolerance in rice (Wang et al., 2016). This represents just the tip of the iceberg, highlighting the vast potential of epigenetic engineering for advancing agricultural practices. Molecular breeders can apply epigenetic techniques to enhance potentially any plant trait, including crop yield, quality, and nutritional value. Only as an example, researchers have identified epigenetic alterations associated with fruit ripening in tomatoes (Manning et al., 2006). Epigenetics thus can play a significant role in contributing to global food security, an issue particularly important in the context of changing climate. Through the targeted manipulation of gene expression, epigenetic modifications can be harnessed to the development of stress-tolerant and high-yielding, quality crops, offering a vital tool for addressing the challenges of a growing population.

Conclusion

Climate fluctuations poses a significant threat to agriculture, with intensifying extreme weather events, rising temperatures, and altered precipitation patterns placing immense stress on crop production. Over the course of evolution, epigenetic mechanisms have enabled plants to adapt to these challenges by dynamically regulating their response pathways. Consequently, epigenetics drives evolutionary pathways that play critical roles in shaping crop development and adaptation. By understanding and harnessing the power of epigenetics, we can unlock the potential to develop resilient crop varieties capable of thriving in the face of climate change.

References

Alvarez, M., Bleich, A., & Donohue, K. (2020). Genotypic variation in the persistence of transgenerational responses to seasonal cues. *Evolution, 74*(10), 2265–2280.

Birnbaum, K. D., & Roudier, F. (2017). Epigenetic memory and cell fate reprogramming in plants. *Regeneration, 4*(1), 15–20.

Burn, J. E., Bagnall, D. J., Metzger, J. D., Dennis, E. S., & Peacock, W. J. (1993). DNA methylation, vernalization, and the initiation of flowering. *Proceedings of the National Academy of Sciences, 90*(1), 287–291.

Chang, Y. N., Zhu, C., Jiang, J., Zhang, H., Zhu, J. K., & Duan, C. G. (2020). Epigenetic regulation in plant abiotic stress responses. *Journal of Integrative Plant Biology, 62*(5), 563–580.

Douhovnikoff, V., & Dodd, R. S. (2015). Epigenetics: A potential mechanism for clonal plant success. *Plant Ecology, 216*, 227–233.

Feng, S., Jacobsen, S. E., & Reik, W. (2010). Epigenetic reprogramming in plant and animal development. *Science, 330*(6004), 622–627.

Finnegan, E. J., Robertson, M., & Helliwell, C. A. (2021). Resetting FLOWERING LOCUS C expression after vernalization is just activation in the early embryo by a different name. *Frontiers in Plant Science, 11*, 620155.

Garg, R., Narayana Chevala, V. V. S., Shankar, R., & Jain, M. (2015). Divergent DNA methylation patterns associated with gene expression in rice cultivars with contrasting drought and salinity stress response. *Scientific Reports, 5*(1), 14922.

Gehring, M. (2019). Epigenetic dynamics during flowering plant reproduction: Evidence for reprogramming? *New Phytologist, 224*(1), 91–96.

Gupta, C., & Salgotra, R. K. (2022). Epigenetics and its role in effecting agronomical traits. *Frontiers in Plant Science, 13*, 925688.

Gutierrez-Marcos, J. F., & Dickinson, H. G. (2012). Epigenetic reprogramming in plant reproductive lineages. *Plant and Cell Physiology, 53*(5), 817–823.

Henderson-Carter, A. L., Kinmonth-Schultz, H., Hileman, L., & Ward, J. K. (2023). FLOWERING LOCUS C drives delayed flowering in Arabidopsis grown and selected at elevated CO2. *bioRxiv.* https://doi.org/10.1101/2023.06.15.545149

Jacobsen, S. E., & Meyerowitz, E. M. (1997). Hypermethylated superman epigenetic alleles in Arabidopsis. *Science, 277*, 1100–1103.

Jacobsen, S. E., Sakai, H., Finnegan, E. J., Cao, X., & Meyerowitz, E. M. (2000). Ectopic hypermethylation of flower specific genes in Arabidopsis. *Current Biology, 24*, 179–186.

Kinoshita, T., & Seki, M. (2014). Epigenetic memory for stress response and adaptation in plants. *Plant and Cell Physiology, 55*(11), 1859–1863.

Manning, K., Tör, M., Poole, M., Hong, Y., Thompson, A. J., King, G. J., et al. (2006). A naturally occurring epigenetic mutation in a gene encoding an SBP-box transcription factor inhibits tomato fruit ripening. *Nature Genetics, 38*(8), 948–952.

Michaels, S. D., & Amasino, R. M. (1999). FLOWERING LOCUS C encodes a novel MADS domain protein that acts as a repressor of flowering. *The Plant Cell, 11*(5), 949–956.

Miura, K., Agetsuma, M., Kitano, H., Yoshimura, A., Matsuoka, M., Jacobsen, S. E., & Ashikari, M. (2009). A metastable dwarf1 epigenetic mutant affecting plant stature in rice. *Proc. Natl. Acad. Sci. USA, 106*, 11218–11223.

Morao, A. K., Bouyer, D., & Roudier, F. (2016). Emerging concepts in chromatin-level regulation of plant cell differentiation: Timing, counting, sensing and maintaining. *Current Opinion in Plant Biology, 34*, 27–34.

Mounger, J., Ainouche, M. L., Bossdorf, O., Cavé-Radet, A., Li, B., Parepa, M., et al. (2021). Epigenetics and the success of invasive plants. *Philosophical Transactions of the Royal Society B, 376*(1826), 20200117.

Paszkowski, J., & Grossniklaus, U. (2011). Selected aspects of transgenerational epigenetic inheritance and resetting in plants. *Current Opinion in Plant Biology, 14*(2), 195–203.

Paun, O., Fay, M. F., Soltis, D. E., & Chase, M. W. (2007). Genetic and epigenetic alterations after hybridization and genome doubling. *Taxon, 56*(3), 649–656.

Pimpinelli, S., & Piacentini, L. (2020). Environmental change and the evolution of genomes: Transposable elements as translators of phenotypic plasticity into genotypic variability. *Functional Ecology, 34*(2), 428–441.

Quadrana, L., & Colot, V. (2016). Plant transgenerational epigenetics. *Annual Review of Genetics, 50*(1), 467–491.

Rajpal, V. R., Rathore, P., Mehta, S., Wadhwa, N., Yadav, P., Berry, E., et al. (2022). Epigenetic variation: A major player in facilitating plant fitness under changing environmental conditions. *Frontiers in Cell and Developmental Biology, 10*, 1020958.

Rehman, M., & Tanti, B. (2020). Understanding epigenetic modifications in response to abiotic stresses in plants. *Biocatalysis and Agricultural Biotechnology, 27*, 101673.

She, W., Grimanelli, D., Rutowicz, K., Whitehead, M. W., Puzio, M., Kotliński, M., et al. (2013). Chromatin reprogramming during the somatic-to-reproductive cell fate transition in plants. *Development, 140*(19), 4008–4019.

Shi, M., Wang, C., Wang, P., Yun, F., Liu, Z., Ye, F., et al. (2023). Role of methylation in vernalization and photoperiod pathway: A potential flowering regulator? *Horticulture Research, 10*(10), uhad174.

Stokes, T. L., Kunkel, B. N., & Richards, E. J. (2002). Epigenetic variation in Arabidopsis disease resistance. *Genes & Development, 16*, 171–182.

Tao, Z., Shen, L., Gu, X., Wang, Y., Yu, H., & He, Y. (2017). Embryonic epigenetic reprogramming by a pioneer transcription factor in plants. *Nature, 551*(7678), 124–128.

Thiebaut, F., Hemerly, A. S., & Ferreira, P. C. G. (2019). A role for epigenetic regulation in the adaptation and stress responses of non-model plants. *Frontiers in Plant Science, 10*, 246.

Vaschetto, L. M. (2015). Exploring an emerging issue: Crop epigenetics. *Plant Molecular Biology Reporter*. Springer New York Publishing Company, *33*, 751–755.

Wang, W., Qin, Q., Sun, F., Wang, Y., Xu, D., Li, Z., & Fu, B. (2016). Genome-wide differences in DNA methylation changes in two contrasting rice genotypes in response to drought conditions. *Frontiers in Plant Science, 7*, 1675.

Wang, Q., Xu, J., Pu, X., Lv, H., Liu, Y., Ma, H., et al. (2021). Maize DNA methylation in response to drought stress is involved in target gene expression and alternative splicing. *International Journal of Molecular Sciences, 22*(15), 8285.

Zhang, L., Cheng, Z., Qin, R., Qiu, Y., Wang, J. L., Cui, X., Gu, L., Zhang, X., Guo, X., Wang, D., et al. (2012). Identification and characterization of an epi-allele of fie1 reveals a regulatory linkage between two epigenetic marks in rice. *Plant Cell, 24*, 4407–4421.

Part II
Epigenetic Mechanisms: Rethinking Soft Inheritance

Chapter 2
DNA Methylation, Histone Modifications, and Non-coding RNA Pathways

The traditional view of inheritance dictated by DNA sequence has already been challenged by epigenetics. The new perspective suggests a "soft inheritance," where environmental cues and developmental programs may leave lasting impressions on the epigenetic landscape through covalent modifications like DNA methylation and histone modifications, and the transmission of non-coding RNAs. These epigenetic marks can activate or deactivate in response to environment and persist across generations, influencing offspring phenotypes without altering the DNA sequence. Hence, epigenetic inheritance allows to reversibly adapt to changing environments without the need of waiting for the selection of genetic mutations, ultimately providing a versatile layer of flexibility in evolutionary fitness. By understanding epigenetic inheritance mechanisms, we can gain invaluable insights into crop development, stress responses, and molecular breeding.

DNA Methylation

DNA methylation involves the enzymatic addition of a methyl group (-CH3) to the cytosine (C) base of DNA. Cytosine represents one of the four fundamental building blocks of DNA, alongside adenine (A), guanine (G), and thymine (T). Methylated genes are generally less likely to be transcribed, and DNA methylation can also affect the splicing of RNA transcripts, leading to different protein isoforms being produced. In plants, DNA methylation exhibits unique patterns and pathways compared to those observed in animals and fungi. This unique characteristic involves the methylation of cytosines in different sequence contexts, specifically CG, CHG, and CHH (where H is A, C, or T), with each context maintained by specific enzymes. Additionally, plants have evolved a diverse set of DNA methylation readers, proteins that specifically recognize and interact with methylated sites within the genome, further emphasizing the complex regulatory network involved.

L. M. Vaschetto, *Epigenetics in Crop Improvement*,
https://doi.org/10.1007/978-3-031-73176-1_2

Plant DNA methylation patterns at specific nucleotide sequence contexts are established by the DNA (cytosine-5)-methyltransferase DRM2 through the RNA-directed DNA methylation (RdDM) pathway. However, maintenance differs: CG methylation relies on the methyltransferase 1 (MET1) and its cofactor VIM, CHG methylation is primarily maintained by the plant-specific Chromomethylase 3 (CMT3) enzyme, and CHH methylation utilizes both the RdDM pathway with DRM2 and Chromomethylase 2 (CMT2)-mediated methylation (Leichter et al., 2022). The presence of specific DNA sequence motifs is required to recruit distinct DNA methyltransferases (DNMTs) and other regulatory factors, thereby establishing or maintaining DNA methylation at specific genomic *loci*.

The plant landscape of DNA methylation exhibits remarkable dynamism, with the degree of methylation varying significantly across different tissue types and developmental stages. DNA methylation patterns are shaped by a variety of factors, including nutrient uptake and stress responses. For instance, exposure to cold temperatures can alter DNA methylation, potentially modifying gene expression and serving as an adaptive response to environmental stress (Gutschker et al., 2022). Exposure to heat can also alter the DNA methylation profile, ultimately promoting cellular plasticity and developmental processes (Sun et al., 2022). Moreover, specific nutrients also play key roles in regulating DNA methylation levels, highlighting the intricate interplay between nutrition and epigenetic modifications in higher organisms (Fan et al., 2022). The dynamic balance between DNA methylation and demethylation is maintained by a diverse array of enzymes, including DNA methyltransferases (DNMTs) and DNA demethylases (dMTases), respectively, which act as molecular architects that add and remove methyl groups to the DNA molecule and thus shape gene expression.

The genomic context also plays a pivotal role in dictating DNA methylation. For instance, promoter gene regions exhibit increased levels of DNA methylation under stress conditions, ultimately suppressing gene expression (Bilichak et al., 2012). Similarly, gene bodies are frequently methylated, contributing to transcriptional repression. Repetitive elements, such as those derived from transposable element insertions, tend to be heavily methylated to suppress their activity and maintain genome stability. Additionally, DNA nucleotide motifs exert a profound influence on DNA methylation profiles, with certain sequence contexts exhibiting an increased susceptibility to DNA methylation.

Histone Modifications

Histone modifications, including acetylation, methylation, phosphorylation, and ubiquitination, represent critical layers of epigenetic regulation that profoundly influence chromatin structure and gene expression. Histone acetylation, the addition of acetyl groups (-COCH3) to lysine residues, is a well-known epigenetic modification capable of neutralizing their positive charges, leading to a reduction in electrostatic interactions between histones and DNA. This ultimately leads to a more relaxed chromatin structure and promotes gene expression. Histone deacetylases (HDACs) counterbalance this process by removing acetyl groups, contributing to gene silencing through the condensation of chromatin (Tian et al., 2005).

Histone methylation, the addition of methyl groups to lysine or arginine residues, is another well-established epigenetic modification that can have either activating or repressive effects on gene expression, depending on the specific residue targeted and the number of methyl groups added (He et al., 2021). Histone methyltransferases (HMTs) are responsible for catalyzing the addition of methyl groups, while histone demethylases (KDMs) remove them, enabling the dynamic regulation of gene expression.

Beyond acetylation and methylation, histone phosphorylation and ubiquitination also play critical roles in regulating chromatin structure and gene expression. Histone kinases catalyze the addition of phosphate groups to serine, threonine, or tyrosine residues, altering chromatin compaction and accessibility. In plant cells, protein-serine/threonine kinases work like a control center, receiving signals from the environment (Hardie, 1999), ultimately turning them into actions such as gene activation and consequent growth. Conversely, histone phosphatases remove phosphate groups, contributing to dynamic regulation of gene expression. The specific residues targeted and the context of the chromatin region determine whether phosphorylation promotes transcriptional activation or repression. Additionally, histone ubiquitination, the covalent attachment of ubiquitin molecules to lysine residues, exerts diverse effects on gene expression. The specific ubiquitin ligases involved determine the consequences of ubiquitination, ranging from influencing chromatin structure and recruitment of chromatin-modifying complexes to modulating protein stability and degradation.

Functions and Regulation of Histone Modifications

Histone modifications represent a critical layer of epigenetic control, shaping chromatin structure and gene expression with profound implications for plant development, stress responses, and epigenetic inheritance. Remarkably, histone acetylation and methylation exhibit a close relationship with transcriptional regulation. As mentioned before, histone acetylation, generally associated with open chromatin configurations, promotes active gene transcription, while deacetylation leads to chromatin condensation and gene repression. Conversely, histone methylation exerts either activating or repressive effects on gene expression, depending on the specific lysine residue targeted and the number of methyl groups added.

Recent studies have shown that changes to histone molecules play a key role in how plants grow and respond to their environment. These changes act like a memory, allowing plants to adjust to future environmental challenges (Zhao et al., 2019). Furthermore, a complex crosstalk exists between various histone modifications, such as the synergistic interplay between histone acetylation and methylation, which exerts additive effects on gene regulation. Histone modifications serve as recognition motifs for protein binding, influencing the recruitment and activity of specific proteins and chromatin-modifying complexes, thereby orchestrating a network of epigenetic regulation (Hoppmann et al., 2011). The regulation of histone modifications involves the intricate interplay of multiple factors, including histone-modifying enzymes, chromatin remodelers, transcription factors, and environmental cues. Other epigenetic pathways, such as the RNA-directed DNA methylation (RdDM) pathway, which requires the activity of the plant-specific DNA-dependent RNA polymerases, can also influence the histone mark landscape (Matzke & Mosher, 2014).

Non-coding RNA Pathways

Epigenetic pathways extend beyond DNA methylation and histone modifications, encompassing a diverse repertoire of regulatory non-coding RNA (ncRNA) molecules capable of modulating gene expression and performing critical biological functions. These non-coding sequences have emerged in the last two decades as critical players in multiple aspects of plant physiology, including development, stress responses, and defense against pathogens. NcRNAs orchestrate key developmental processes such as leaf morphogenesis, root architecture, and floral patterning, thereby contributing to the formation and function of plant organs. Additionally, ncRNAs play pivotal roles in stress responses like drought, salinity, and pathogen infections, regulating the expression of stress-responsive genes and enabling plants to trigger adaptive responses and face environmental challenges (Yu et al., 2019). NcRNAs contribute to the defense against pathogens by targeting viral or pathogen-derived transcripts for degradation or by regulating the expression of defense-related genes, ultimately enhancing plant resilience against diseases (Samarfard et al., 2022).

Classification and Biogenesis of ncRNAs

NcRNAs can be classified according to their size into long ncRNAs (lncRNAs, >200 nt) and short ncRNAs (sncRNAs, <200 nt). Based on their biogenesis and functional mechanisms, sncRNAs are further categorized into distinct classes. Two of the most important types of sncRNAs include microRNAs (miRNAs) and small interfering RNAs (siRNAs). miRNAs are a class of evolutionarily conserved ncRNAs with approximately 21 nucleotides in length, derived from longer primary transcripts (pri-miRNAs). These pri-miRNAs are processed by an enzyme called DICER-LIKE1 (DCL1) into precursor miRNAs (pre-miRNAs), which are further matured into functional miRNAs. Mature miRNAs guide the RNA-induced silencing complex (RISC) to target messenger RNAs (mRNAs), ultimately leading to mRNA degradation or translational repression. Small interfering RNAs (siRNAs) represent another class of small ncRNAs involved in plant gene regulation. These siRNAs are derived from double-stranded RNA (dsRNA) precursors, which can originate from diverse sources like repetitive elements, transgenes, and viral RNA. DCL enzymes play a critical role in processing dsRNAs into siRNA duplexes, which are subsequently loaded into ARGONAUTE (Ago) proteins within the RISC complex. The siRNA-guided RISC complex is then directed towards complementary target RNAs, leading to either mRNA degradation or transcriptional gene silencing (TGS), highlighting the diverse mechanisms employed by siRNAs to orchestrate gene expression in plants.

Long non-coding RNAs (lncRNAs) are another important class of ncRNAs with diverse functions and significant impact on gene expression. These sequences can be transcribed from intergenic regions or overlap with exons of protein-coding genes, contributing to the complex transcriptional landscape. Based on their biogenesis, lncRNAs can be further categorized into intergenic lncRNAs (lincRNAs), intronic lncRNAs (incRNAs), natural antisense lncRNAs (NAT-lncRNAs, arising from opposite strands of protein-coding genes), and circular lncRNAs (circRNAs, formed through back-splicing of pre-mRNA exons) (Jha et al., 2020). These sequences participate in a wide range of regulatory processes, including transcriptional activation and repression, alternative splicing, and mRNA stability, thereby influencing gene expression at multiple levels. Remarkably, their diverse modes of action involve functioning as scaffolds, decoys, sponges, or guides for specific protein complexes including histone acetyltransferases (HATs) and histone deacetylases (HDACs), or RNA-binding proteins. Consequently, lncRNA shapes gene expression and chromatin dynamics through diverse mechanisms, including histone modifications, RNA degradation pathways, and chromatin remodeling.

Circular RNAs (circRNAs), a widespread class of endogenous lncRNAs, deserve special mention. Over the past few years, research on circRNAs has shed light on their potential significance across diverse higher organisms ranging from animals to plants. The ubiquitous presence of circRNAs across the plant kingdom, and their differential expression during stress and developmental phases, evidence important roles in these processes. Circular RNAs (circRNAs) are critical regulators of gene expression under stress conditions in both monocot and dicot crops (Vaschetto et al., 2020; Mannan et al., 2022). Plant circRNAs exhibit a wide range of sizes, which can vary depending on the species and tissue type. Through a unique splicing event termed 'backsplicing,' circRNAs arise when a downstream 5′ splice site joins covalently with an upstream 3′ splice site, forming a closed-loop RNA molecule. These molecules exhibit diverse gene regulatory functions, including sponging microRNAs, competing with endogenous RNAs, and influencing alternative mRNA splicing.

NcRNAs and Epigenetic Pathways

NcRNAs exert a profound influence on gene regulation across various levels, orchestrating complex epigenetic networks. Their diverse functions encompass gene silencing, post-transcriptional regulation, and epigenetic modifications (Zhang et al., 2013; Tenea, 2009; Rajwanshi et al., 2014; Heo et al., 2013). This intricate mechanism of regulation is crucial for various developmental processes, including organogenesis, flowering time control, and hormone signaling, highlighting the diverse and essential roles of ncRNAs in shaping plant physiology. In particular, ncRNAs participate in various epigenetic modifications, including the establishment and maintenance of DNA methylation and histone modifications. For example, siRNAs derived from transposable elements or repetitive DNA sequences guide RNA-directed DNA methylation (RdDM) pathways, leading to DNA methylation and heterochromatin formation (Rigal & Mathieu, 2011; Ariel & Manavella, 2021). Furthermore, lncRNAs are also crucial players in the association between gene expression and chromatin organization. They can interact with chromatin modifiers, recruiting them to specific genomic *loci* to regulate tridimensional (3D) chromatin structure. Consequently, lncRNAs can directly interact with chromatin via specific DNA sequences or protein-protein interactions, serving as adaptors to recruit chromatin modifiers, such as histone acetyltransferases (HATs) and histone deacetylases (HDACs), to specific target *loci*. This targeted recruitment facilitates localized changes in histone modifications and DNA methylation patterns, contributing to the dynamic regulation of gene expression. Moreover, specific lncRNAs can function as scaffolds for the assembly of chromatin-modifying complexes, facilitating their recruitment and promoting localized chromatin modifications. Finally, certain lncRNAs can also act as decoys, sequestering chromatin modifiers and preventing their access to specific genomic regions, thereby contributing to the dynamic regulation of chromatin dynamics and gene expression (Chen et al., 2021).

DNA methylation and histone modifications often work in concert to orchestrate chromatin structure and influence gene expression patterns. DNA methylation at gene promoters can trigger the recruitment of histone deacetylases (HDACs) and histone methyltransferases (HMTs), leading to gene repression through histone deacetylation and methylation. Reciprocally, histone modifications can also exert influence on DNA methylation patterns. Histone modifications, such as acetylation, can promote DNA demethylation through the recruitment of DNA demethylases or by preventing the binding of DNA methyltransferases. This intricate interplay between histone modifications and DNA methylation contributes to the dynamic regulation of gene expression and chromatin states.

The interplay between DNA methylation, histone modifications, and non-coding RNAs (ncRNAs) is often characterized by dynamic feedback loops and interconnected regulatory networks (Ramirez-Prado et al., 2017). For instance, DNA methylation can directly influence the expression of ncRNAs either through the recruitment of DNA methyltransferases or the modification of chromatin structure. These ncRNAs, in turn, guide DNA methylation or histone modifications to specific genomic *loci*, ultimately contributing to the establishment and maintenance of stable epigenetic states (Avramova, 2011). Additionally, histone modifications can also exert diverse regulatory effects on ncRNAs, influencing their expression, processing, and stability. This complex interplay results in the further modulation of chromatin states and contributes to the dynamic regulation of gene expression.

Chromatin-Remodeling Complexes

Chromatin-remodeling complexes (CRCs) are ATP-dependent molecular structures that orchestrate gene expression and various DNA-related processes by reorganizing the chromatin, modulating histone-DNA interactions, and enabling specific protein binding to nucleosomal DNA. These ATP-dependent complexes are evolutionarily conserved among different eukaryotes, including plants. Plant CRCs, which include SWI/SNF, ISWI, INO80, and CHD, have been shown to control genome integrity, gene transcription, and modulate responses to environmental stresses (Yang et al., 2022; Wang et al., 2019; Ojolo et al., 2018). They control the distribution pattern and density of nucleosomes, influencing the accessibility of DNA for gene expression. Remarkably, CRCs are crucial players that promote histone modifications, consequently altering the 3D chromatin structure and regulating gene expression.

SWI/SNF

The SWI/SNF (switch/sucrose nonfermenting) complex is an evolutionary conserved subfamily of ATP-dependent CRCs that form large protein assemblies essential for regulating gene expression in plants. They consist of multiple subunits, including a central ATPase enzyme from the Snf2 family. In *Arabidopsis*, three well-characterized SWI/SNF CRCs share these core components: a Snf2 family ATPase (BRM, SYD, or MINU1/2), two actin-related proteins (ARP4 and ARP7), and various subunits forming the base module (SWI3A-D, SWP73A/B, BSH, and AN3) (Bieluszewski et al., 2023). These SWI/SNF CRCs play critical roles in gene regulation through diverse mechanisms:

1. Enhancing transcription factor accessibility: SWI/SNF CRCs facilitate the binding of transcription factors (TFs) to their cognate sites within enhancers, thereby promoting gene expression.
2. Overcoming Polycomb-mediated silencing: SWI/SNF CRCs can counteract the repressive effects of Polycomb group (PcG) proteins, which silence gene expression through epigenetic modifications.
3. Regulation of chromatin looping: It has been shown that the *Arabidopsis* genome is organized into compartments and includes *locus*-specific loops formed by

contacts between gene body ends (5′ and 3′). These loops enhance interactions between enhancers and promoters, ultimately influencing gene expression (Bieluszewski et al., 2023). Interestingly, in *Arabidopsis*, it has been shown that SWI/SNF CRCs can prevent the formation of specific activating loops at the *Flowering Locus C (FLC) locus* (Jégu et al., 2014). FLC is the central regulator of the induction of flowering by vernalization, promoting flowering in late-flowering ecotypes and mutants (Sheldon et al., 2000). In a climate change scenario with unpredictable cold spells, maintaining vernalization requirements through the SWI/SNF function can be beneficial by potentially preventing plants from flowering prematurely during unseasonal warm periods, ultimately leading to increased plant resilience.

In rice, Growth-Regulating Factors (GRFs) function within the gibberellin (GA) signaling pathway. These GRFs interact with GRF-Interacting Factors (GIFs), which in turn recruit the SWI/SNF chromatin remodeling complex. DELLA proteins, key transcription factors during plant development, regulate GRF levels by influencing both miR396 levels, a miRNA responsive to multiple developmental and environmental cues, and directly interacting with the GRFs themselves (Lu et al., 2022). Rice also contains a protein called OsLFR, identified as an ortholog of the *Arabidopsis* LFR protein, which is a SWI/SNF CRC component (Qi et al., 2020). Protein-protein interaction analyses revealed that OsLFR physically interacts with several other predicted components of the rice SWI/SNF CRC. Furthermore, OsLFR plays an essential role in regulating rice seed development (Qi et al., 2020). In maize, Yu et al. (2016) identified ZmCHB101, a SWI3 subunit, as a key regulator of growth and development. This subunit orchestrates the expression of a broad spectrum of genes governing metabolic processes, photosynthesis, stress response, and transcriptional regulation. Remarkably, ZmCHB101 also plays a critical role in maintaining normal nucleosome density and 45S rDNA compaction (Yu et al., 2016). SWI/SNF CRC components also interacts with lncRNAs to modulate epigenetic pathways. In *Arabidopsis*, SWI3B works with the IDN2/IDP complex to target specific regions of DNA for methylation. This IDN2/IDP complex binds to lncRNAs produced by a specialized RNA polymerase (Pol V). By influencing nucleosome positioning near these Pol V sites, SWI3B facilitates RNA-directed DNA methylation, which helps silence transposable elements (Zhang et al., 2018).

ISWI

Members of the Imitation Switch (ISWI) subfamily represent another class of ATP-dependent chromatin remodeling factors. Consequently, these proteins also utilize ATP hydrolysis to reposition nucleosomes, thereby regulating DNA accessibility (Li et al., 2017). In *Arabidopsis*, ISWI is essential for establishing the distributionally spaced nucleosome pattern within gene bodies, a hallmark of highly transcribed genes (Li et al., 2017). Beyond this core function, ISWI family members contribute to diverse cellular processes and phenotypic outcomes in plants, including stamen development and fertility (Gu et al., 2020; Zhao et al., 2021), immune responses (Liu et al., 2021), floral transition (Noh & Amasino, 2003; Tan et al., 2020; Corcoran et al., 2022), leaf development (Han et al., 2018), and vegetative growth (Li et al., 2012). In *Arabidopsis*, the ISWI chromatin remodeling protein PHOTOPERIOD-INDEPENDENT EARLY FLOWERING1 (PIE1) is essential for FLC activation and floral repression. In this plant species, PIE1 activity facilitates *FLC* expression to levels sufficient for inhibiting flowering (Noh & Amasino, 2003). Tan et al. (2020) revealed that the ARID5 subunit of the plant-specific ISWI complex regulates development and floral transition. The ARID-PHD domain of ARID5 recognizes both AT-rich DNA and H3K4me3 marks, facilitating the recruitment of the ISWI complex to specific chromatin regions to influence development and floral transition. Corcoran et al. (2022) revealed that specific residues within histone H4 influence nucleosome spacing via their interaction with ISWI. Interestingly, the authors also observed that these H4 modifications promote earlier flowering. Moreover, Gu et al. (2020) identified a plant-specific subunit of the *Arabidopsis* ISWI complex, FHA2, that directly binds to RLT1 and RLT2, two functionally redundant subunits of the same complex, ultimately governing stamen development and plant fertility. Li et al. (2012) demonstrated that the ISWI members CHR11 and CHR17 interact with DDT-domain proteins, thereby preventing plants from initiating the vegetative-to-reproductive transition prematurely. CHR11 and CHR17 achieve this effect by regulating several flowering genes that control flower timing, while also promoting *FLC* expression in leaves.

INO80

Initially identified as a transcriptional co-activator in yeast inositol metabolism (Ebbert et al., 1999), INO80 is another evolutionarily conserved subfamily of chromatin remodeling complexes (CRCs) in eukaryotes. It functions as a scaffold to assemble other proteins into the INO80 complex, which has been implicated in transcriptional activation via nucleosome eviction, a process that leads to the removal or displacement of nucleosomes, increasing the accessibility of DNA sequences. While studied in various eukaryotes, plant INO80 functions have only been explored in *Arabidopsis* until recently (Zhang et al., 2015). For instance, it has been shown that the *Arabidopsis* INO80 protein (AtINO80) controls homologous recombination (Fritsch et al., 2004), and also plays crucial roles in maintaining genome stability and plant development (Zhang et al., 2015; Kang et al., 2019). A key aspect of plant responses to climate change is thermomorphogenesis, a suite of morphological changes plants undergo in response to altered ambient temperatures. Remarkably, AtINO80 has been shown to promote thermomorphogenesis by coupling the eviction of the histone variant H2A.Z from nucleosomes with active gene transcription (Xue et al., 2021). Shang et al. (2021) revealed that histone H3K4 methyltransferase complexes interact with the N-terminal domain of AtINO80, facilitating H3K4 trimethylation and promoting transcriptional activation. This study also highlights the opposing roles of the INO80 complex in regulating transcription, which are essential for maintaining a balance between vegetative growth and flowering under various environmental conditions (Shang et al., 2021).

CHD

Chromodomain helicase DNA-binding (CHD) proteins are a subfamily of ATP-dependent CRCs found in plants and animals. These proteins regulate diverse aspects of gene expression (Lu et al., 2020). In *Arabidopsis*, PICKLE, a CHD protein similar to animal subfamily II remodelers, plays a crucial role in repressing genes that control developmental identity (Ho et al., 2013). Deans et al. (2020) identified a maize protein, CHD3 (chromodomain helicase DNA-binding 3), as an ortholog of the *Arabidopsis* PICKLE protein. This CHD3 protein acts as a key component for maintaining both paramutation-associated silencing of the *purple plant1* (*pl1*) gene and normal plant development, including somatic development and male gametophyte function. Intriguingly, the authors identified components involved in 24-nucleotide RNA biogenesis, including the largest subunit of RNA polymerase IV, as essential for maintaining repression of the *pl1* allele. These findings suggest a link between 24-nucleotide RNA biogenesis, RNA polymerase IV activity, and nucleosome positioning in the epigenetic phenomenon of paramutation (Deans et al., 2020). Moreover, in rice, Lu et al. (2020) demonstrated that the CHD protein CHR729 influences nucleosome occupancy, gene expression, and H3K4me3 modifications, primarily affecting underexpressed genes. Also in rice, it has been shown that the CRL6 CHD protein is required for crown root development (Wang et al., 2016). Shen et al. (2015) further revealed a role for the *Arabidopsis* CHD1-like protein CHR5 in regulating seed development. CHR5 expression is detected during embryo development and seed maturation, where it directly activates the expression of key regulators ABI3 and FUS3. Remarkably, the authors propose that CHR5 reduces the nucleosomal barrier, thereby promoting an active state for these essential embryo regulatory genes. This finding suggests potential avenues for manipulating CHR5 activity to enhance seed protein production (Shen et al., 2015).

Conclusions

Understanding the intricate interplay between DNA methylation, histone modifications, and non-coding RNA pathways in plant responses to climate change offers tremendous potential for crop improvement. By targeting these pathways, we can potentially develop climate-resilient crops with enhanced tolerance to climate stressful conditions. Current research focused on deciphering the specific roles of these epigenetic regulators in stress responses and identifying key regulatory elements is crucial for translating this knowledge into practical applications. This surely will pave the way for the development of superior crop varieties, ultimately ensuring food security in challenging climatic conditions.

References

Ariel, F. D., & Manavella, P. A. (2021). When junk DNA turns functional: Transposon-derived non-coding RNAs in plants. *Journal of Experimental Botany, 72*(11), 4132–4143.

Avramova, Z. (2011). Epigenetic regulatory mechanisms in plants. In *Handbook of epigenetics* (pp. 251–278). Academic.

Bieluszewski, T., Prakash, S., Roulé, T., & Wagner, D. (2023). The role and activity of SWI/SNF chromatin remodelers. *Annual Review of Plant Biology, 74*, 139–163.

Bilichak, A., Ilnystkyy, Y., Hollunder, J., & Kovalchuk, I. (2012). The progeny of Arabidopsis thaliana plants exposed to salt exhibit changes in DNA methylation, histone modifications and gene expression. *PLoS One, 7*(1), e30515.

Chen, Q., et al. (2021). From "dark matter" to "star": Insight into the regulation mechanisms of plant functional long non-coding RNAs. *Frontiers in Plant Science, 12*, 650926.

Corcoran, E. T., LeBlanc, C., Huang, Y. C., Arias Tsang, M., Sarkiss, A., Hu, Y., et al. (2022). Systematic histone H4 replacement in Arabidopsis thaliana reveals a role for H4R17 in regulating flowering time. *The Plant Cell, 34*(10), 3611–3631.

Deans, N. C., Giacopelli, B. J., & Hollick, J. B. (2020). Locus-specific paramutation in Zea mays is maintained by a PICKLE-like chromodomain helicase DNA-binding 3 protein controlling development and male gametophyte function. *PLoS Genetics, 16*(12), e1009243.

Ebbert, R., Birkmann, A., & Schuller, H. J. (1999). The product of the SNF2/SWI2 paralogue INO80 of Saccharomyces cerevisiae required for efficient expression of various yeast structural genes is part of a high-molecular-weight protein complex. *Molecular Microbiology, 32*, 741–751.

Fan, X., Peng, L., & Zhang, Y. (2022). Plant DNA methylation responds to nutrient stress. *Genes, 13*(6), 992.

Fritsch, O., Benvenuto, G., Bowler, C., Molinier, J., & Hohn, B. (2004). The INO80 protein controls homologous recombination in Arabidopsis thaliana. *Molecular Cell, 16*(3), 479–485.

Gu, B. W., Tan, L. M., Zhang, C. J., Hou, X. M., Cai, X. W., Chen, S., & He, X. J. (2020). FHA2 is a plant-specific ISWI subunit responsible for stamen development and plant fertility. *Journal of Integrative Plant Biology, 62*(11), 1703–1716.

Gutschker, S., Corral, J. M., Schmiedl, A., Ludewig, F., Koch, W., Fiedler-Wiechers, K., et al. (2022). Multi-omics data integration reveals link between epigenetic modifications and gene expression in sugar beet (Beta vulgaris subsp. vulgaris) in response to cold. *BMC Genomics, 23*(1), 144.

Han, W., Han, D., He, Z., Hu, H., Wu, Q., Zhang, J., et al. (2018). The SWI/SNF subunit SWI3B regulates IAMT1 expression via chromatin remodeling in Arabidopsis leaf development. *Plant Science, 271*, 127–132.

Hardie, D. G. (1999). Plant protein serine/threonine kinases: Classification and functions. *Annual Review of Plant Biology, 50*(1), 97–131.

He, K., Cao, X., & Deng, X. (2021). Histone methylation in epigenetic regulation and temperature responses. *Current Opinion in Plant Biology, 61*, 102001.

Heo, J. B., Lee, Y. S., & Sung, S. (2013). Epigenetic regulation by long noncoding RNAs in plants. *Chromosome Research, 21*, 685–693.

Ho, K. K., Zhang, H., Golden, B. L., & Ogas, J. (2013). PICKLE is a CHD subfamily II ATP-dependent chromatin remodeling factor. *Biochimica et Biophysica Acta (BBA)-Gene Regulatory Mechanisms, 1829*(2), 199–210.

Hoppmann, V., Thorstensen, T., Kristiansen, P. E., Veiseth, S. V., Rahman, M. A., Finne, K., et al. (2011). The CW domain, a new histone recognition module in chromatin proteins. *The EMBO Journal, 30*(10), 1939–1952.

Jégu, T., Latrasse, D., Delarue, M., Hirt, H., Domenichini, S., Ariel, F., et al. (2014). The BAF60 subunit of the SWI/SNF chromatin-remodeling complex directly controls the formation of a gene loop at FLOWERING LOCUS C in Arabidopsis. *The Plant Cell, 26*(2), 538–551.

Jha, U. C., Nayyar, H., Jha, R., et al. (2020). Long non-coding RNAs: Emerging players regulating plant abiotic stress response and adaptation. *BMC Plant Biology, 20*, 466. https://doi.org/10.1186/s12870-020-02595-x

Kang, H., Zhang, C., An, Z., Shen, W. H., & Zhu, Y. (2019). AtINO80 and AtARP5 physically interact and play common as well as distinct roles in regulating plant growth and development. *New Phytologist, 223*(1), 336–353.

Leichter, S. M., Du, J., & Zhong, X. (2022). Structure and mechanism of plant DNA methyltransferases. In *DNA Methyltransferases-role and function* (pp. 137–157). Springer International Publishing.

Li, G., Zhang, J., Li, J., Yang, Z., Huang, H., & Xu, L. (2012). Imitation switch chromatin remodeling factors and their interacting RINGLET proteins act together in controlling the plant vegetative phase in Arabidopsis. *The Plant Journal, 72*(2), 261–270.

Li, D., Liu, J., Liu, W., et al. (2017). The ISWI remodeler in plants: Protein complexes, biochemical functions, and developmental roles. *Chromosoma, 126*, 365–373. https://doi.org/10.1007/s00412-017-0626-9

Liu, H., Li, J., Xu, Y., Hua, J., & Zou, B. (2021). ISWI chromatin remodeling factors repress PAD4-mediated plant immune responses in Arabidopsis. *Biochemical and Biophysical Research Communications, 583*, 63–70.

Lu, Y., Tan, F., Zhao, Y., Zhou, S., Chen, X., Hu, Y., & Zhou, D. X. (2020). A chromodomain-helicase-DNA-binding factor functions in chromatin modification and gene regulation. *Plant Physiology, 183*(3), 1035–1046.

Lu, Y., Zeng, J., & Liu, Q. (2022). The rice miR396-GRF-GIF-SWI/SNF module: A player in GA signaling. *Frontiers in Plant Science, 12*, 786641.

Mannan, M., Rima, I., & Karim, A. (2022). Physiological and biochemical basis of stress tolerance in soybean. In *Soybean-recent advances in research and applications*. IntechOpen.

Matzke, M. A., & Mosher, R. A. (2014). RNA-directed DNA methylation: An epigenetic pathway of increasing complexity. *Nature Reviews Genetics, 15*(6), 394–408.

Noh, Y. S., & Amasino, R. M. (2003). PIE1, an ISWI family gene, is required for FLC activation and floral repression in Arabidopsis. *The Plant Cell, 15*(7), 1671–1682.

Ojolo, S. P., Cao, S., Priyadarshani, S. V. G. N., Li, W., Yan, M., Aslam, M., et al. (2018). Regulation of plant growth and development: A review from a chromatin remodeling perspective. *Frontiers in Plant Science, 9*, 1232.

Qi, D., Wen, Q., Meng, Z., Yuan, S., Guo, H., Zhao, H., & Cui, S. (2020). OsLFR is essential for early endosperm and embryo development by interacting with SWI/SNF complex members in Oryza sativa. *The Plant Journal, 104*(4), 901–916.

Rajwanshi, R., Chakraborty, S., Jayanandi, K., Deb, B., & Lightfoot, D. A. (2014). Orthologous plant microRNAs: Microregulators with great potential for improving stress tolerance in plants. *Theoretical and Applied Genetics, 127*, 2525–2543.

Ramirez-Prado, J. S., Ariel, F., Benhamed, M., & Crespi, M. (2017). Plant epigenetics: Non-coding RNAs as emerging regulators. In N. Rajewsky, S. Jurga, & J. Barciszewski (Eds.), *Plant Epigenetics. RNA Technologies* (pp. 129–147). Springer.

Rigal, M., & Mathieu, O. (2011). A "mille-feuille" of silencing: Epigenetic control of transposable elements. *Biochimica et Biophysica Acta (BBA)-Gene Regulatory Mechanisms, 1809*(8), 452–458.

Samarfard, S., Ghorbani, A., Karbanowicz, T. P., Lim, Z. X., Saedi, M., Fariborzi, N., et al. (2022). Regulatory non-coding RNA: The core defense mechanism against plant pathogens. *Journal of Biotechnology, 359*, 82–94.

Shang, J. Y., Lu, Y. J., Cai, X. W., Su, Y. N., Feng, C., Li, L., et al. (2021). COMPASS functions as a module of the INO80 chromatin remodeling complex to mediate histone H3K4 methylation in Arabidopsis. *The Plant Cell, 33*(10), 3250–3271.

Sheldon, C. C., Rouse, D. T., Finnegan, E. J., Peacock, W. J., & Dennis, E. S. (2000). The molecular basis of vernalization: The central role of FLOWERING LOCUS C (FLC). *Proceedings of the National Academy of Sciences, 97*(7), 3753–3758.

Shen, Y., Devic, M., Lepiniec, L., & Zhou, D. X. (2015). Chromodomain, helicase and DNA-binding CHD1 protein, CHR5, are involved in establishing active chromatin state of seed maturation genes. *Plant Biotechnology Journal, 13*(6), 811–820.

Sun, M., Yang, Z., Liu, L., & Duan, L. (2022). DNA methylation in plant responses and adaption to abiotic stresses. *International Journal of Molecular Sciences, 23*(13), 6910.

Tan, L. M., Liu, R., Gu, B. W., Zhang, C. J., Luo, J., Guo, J., et al. (2020). Dual recognition of H3K4me3 and DNA by the ISWI component ARID5 regulates the floral transition in Arabidopsis. *The Plant Cell, 32*(7), 2178–2195.

Tenea, G. N. (2009). Exploring the world of RNA interference in plant functional genomics: A research tool for many biology phenomena. *Centre of Microbial Biotechnology, 14*, 4360–4364.

Tian, L., Fong, M. P., Wang, J. J., Wei, N. E., Jiang, H., Doerge, R. W., & Chen, Z. J. (2005). Reversible histone acetylation and deacetylation mediate genome-wide, promoter-dependent and locus-specific changes in gene expression during plant development. *Genetics, 169*(1), 337–345.

Vaschetto, L. M., Litholdo, C. G., Sendín, L. N., Terenti Romero, C. M., & Filippone, M. P. (2020). Cereal circular RNAs (circRNAs): An overview of the computational resources for identification and analysis. *Cereal Genomics: Methods and Protocols, 2072*, 157–163.

Wang, Y., Wang, D., Gan, T., Liu, L., Long, W., Wang, Y., et al. (2016). CRL6, a member of the CHD protein family, is required for crown root development in rice. *Plant Physiology and Biochemistry, 105*, 185–194.

Wang, J., Gao, S., Peng, X., Wu, K., & Yang, S. (2019). Roles of the INO80 and SWR1 chromatin remodeling complexes in plants. *International Journal of Molecular Sciences, 20*(18), 4591.

Xue, M., Zhang, H., Zhao, F., Zhao, T., Li, H., & Jiang, D. (2021). The INO80 chromatin remodeling complex promotes thermomorphogenesis by connecting H2A. Z eviction and active transcription in Arabidopsis. *Molecular Plant, 14*(11), 1799–1813.

Yang, T., Wang, D., Tian, G., Sun, L., Yang, M., Yin, X., et al. (2022). Chromatin remodeling complexes regulate genome architecture in Arabidopsis. *The Plant Cell, 34*(7), 2638–2651.

Yu, X., Jiang, L., Wu, R., Meng, X., Zhang, A., Li, N., et al. (2016). The core subunit of a chromatin-remodeling complex, ZmCHB101, plays essential roles in maize growth and development. *Scientific Reports, 6*(1), 38504.

Yu, Y., Zhang, Y., Chen, X., & Chen, Y. (2019). Plant noncoding RNAs: Hidden players in development and stress responses. *Annual Review of Cell and Developmental Biology, 35*(1), 407–431.

Zhang, J., Mujahid, H., Hou, Y., Nallamilli, B. R., & Peng, Z. (2013). Plant long ncRNAs: A new frontier for gene regulatory control. *American Journal of Plant Sciences, 04*, 1038–1045.

Zhang, C., Cao, L., Rong, L., An, Z., Zhou, W., Ma, J., et al. (2015). The chromatin-remodeling factor at INO 80 plays crucial roles in genome stability maintenance and in plant development. *The Plant Journal, 82*(4), 655–668.

Zhang, H., Lang, Z., & Zhu, J. K. (2018). Dynamics and function of DNA methylation in plants. *Nature Reviews Molecular Cell Biology, 19*(8), 489–506.

Zhao, T., Zhan, Z., & Jiang, D. (2019). Histone modifications and their regulatory roles in plant development and environmental memory. *Journal of Genetics and Genomics, 46*(10), 467–476.

Zhao, Y., Jiang, T., Li, L., Zhang, X., Yang, T., Liu, C., et al. (2021). The chromatin remodeling complex imitation of switch controls stamen filament elongation by promoting jasmonic acid biosynthesis in Arabidopsis. *Journal of Genetics and Genomics, 48*(2), 123–133.

Chapter 3
The Role of Epigenetic Variation in Plant Adaptation

Epigenetic variation exhibits a dual system for achieving adaptation outcomes. While most changes have short-term effects, some have shown to persist across generations, shaping heritable traits. These transient responses provide a first line for flexible adaptability, while stable epigenetic modifications, which can be transmitted across generations, enhance the organism's ability to cope with recurring or prolonged stress. The transmission of epigenetic marks or "epialleles" enriches the organism's adaptive potential. The remarkable stability and heritable nature of some epialleles allow for the transmission of acquired traits across generations.

Epigenetics as Source of Variation

Epigenetic variation plays a pivotal role in stress adaptation within natural populations. When plants confront stress (e.g., drought, extreme temperatures, salinity, or other adverse conditions) they can undergo epigenetic changes. These changes entail the activation of stress-responsive genes or the silencing of non-essential genes, allowing plants to face stress and increase their odds of survival and reproduction. One of the most important aspects of epigenetic variation is its capacity to generate phenotypic diversity within populations. This means that, in response to environmental stress, plants with different epigenetic profiles can exhibit diverse phenotypic responses. Therefore, epigenetic diversity constitutes a reservoir of adaptive traits, with potential to be selected in breeding programs. The strategies focused on selecting and propagating plants with specific epigenetic profiles associated with stress tolerance can lead to the development of more resilient crops.

Transposable elements (TEs) are mobile DNA sequences that insert themselves within genomes and replicate independently of the host cell's DNA. They are major drivers of genome evolution and chromosome rearrangements. In plants, local polymorphisms in TEs, specifically in their insertion sequences, primarily drive

L. M. Vaschetto, *Epigenetics in Crop Improvement*, https://doi.org/10.1007/978-3-031-73176-1_3

epigenetic variation (Baduel & Sasaki, 2023). Understanding the outstanding influence of TE insertions on epigenetic variation, especially in TE-rich plant genomes like maize and wheat, is of paramount importance. Densely methylated TE sequences often alter the epigenetic status of nearby genes and even entire linkage groups, while mutational accumulation within these sequences can also impact their methylation levels (Hollister & Gaut, 2009; Dubin et al., 2015; Quadrana et al., 2016; Vaschetto, 2016).

Transgenerational Stress Adaptation

Despite growing evidence for the ecological significance of transgenerational adaptive responses, the underlying epigenetic mechanisms remain largely unexplored (Weinhold, 2018). Epigenetic inheritance holds significant implications for stress adaptation. It enables a more prompt and efficient response to recurring stressors. For example, if a parent plant experiences drought stress and passes on epigenetic marks associated with stress tolerance to its offspring, the progeny may then gain an advantage when facing new drought conditions. The stress response-related genes might then be pre-activated, and they would exhibit physiological changes that aid in water conservation and better withstand water scarcity.

Transgenerational epigenetic inheritance relies on the faithful transmission of DNA methylation and histone modifications through the germ line, bypassing the resetting mechanisms of gametes and the zygote. This intricate process ensures both genomic stability and the totipotency of zygote cells (Annacondia & Martinez, 2019). Repetitive sequences contribute to the formation of alleles associated with epialleles. Recent evidence suggests that the inheritance of variations in DNA methylation states over these nucleotide sequences may be due to insufficient epigenetic reinforcement, rather than failed reprogramming of DNA methylation (Quadrana & Colot, 2016). Small RNAs also play key roles as carriers of epigenetic information across generations, contributing to the establishment and maintenance of stable epigenetic states. In maize, for instance, it has been shown that transgenerational methylation patterns depend on siRNAs. Cao et al. (2022) observed that siRNAs between hybridizing maize parents can induce differentially methylated epialleles in their hybrid offspring. These induced epigenetic states can be remarkably stable, persisting through hybrid breeding generations, and thus alter gene expression to enhance growth and stress adaptation.

The transmission of epigenetic information through non-coding RNAs (ncRNAs) is not static but rather dynamically influenced by environmental cues and stress conditions. This dynamic interplay between environmental factors, ncRNA expression, and epigenetic states ultimately shapes the transgenerational inheritance of acquired traits. Environmental factors, such as temperature and light, can trigger changes in the expression and activity of miRNAs, siRNAs, and lncRNAs (Zhao et al., 2020; Yu et al., 2019). Moreover, ncRNA-mediated epigenetic modifications may potentially be transmitted to subsequent generations, offering a mechanism for plants to adapt by short-term evolutionary responses to changing environmental conditions.

Transgenerational epigenetic inheritance underscores the concept of environmental plasticity, which means that plants adapt to their environment not solely through genetic changes but also reversible and heritable epigenetic modifications. This plasticity is particularly relevant amid the context of evolution in environmental conditions associated with climate change. Nonetheless, it is important to highlight the frequently observed low stability and persistence of transgenerational epigenetic marks over generations. The specificity and mechanisms governing how certain epigenetic marks are inherited while others are reset during epigenetic reprogramming processes remain as subjects of ongoing research.

Epigenetics, Priming and Stress Memory

Plants exhibit a phenomenon called priming (primarily known as defense priming), where they retain memory of past environmental (e.g., pathogens and/or pests) stress experiences. This memory allows them to mount a more robust response upon encountering the same stress again. Epigenetic modifications in DNA and associated histones underlie this priming, potentially carrying both short-term and long-term memories within the same plant or even mediating transgenerational effects (Harris et al., 2023). Priming is regulated by a complex network of enzymes, including epigenetic enzymes such as DNA methyltransferases and histone methyltransferases, demethylases, acetyltransferases, and deacetylases, ultimately manifesting as the development of changes that enhance responsiveness to subsequent stress encounters. Thus, the "epigenetic priming effect" represents a cellular strategy where the epigenome is purposefully modified to enhance responsiveness to specific environmental cues. Through such a priming effect, plants may potentially retain memory of past environmental stress, leading to a more robust response upon subsequent exposure. This phenomenon serves as a form of stress memory, enabling plants to remember stressful experiences to mount a more rapid and efficient response upon future encounters. The induction of epigenetic stress memory can be triggered by a variety of factors ranging from abiotic to biotic stressors, hormonal signals, and beneficial microbial interactions (Godwin & Farrona, 2020; Turgut-Kara et al., 2020). For instance, pre-exposure to drought and salinity conditions can induce specific epigenetic alterations in stress-responsive genes, leading to enhanced tolerance (Banerjee & Roychoudhury, 2017).

Epigenetic stress memory involves a dynamic interplay between sequence-specific epigenetic modifications at stress-responsive genes and the RNA-directed DNA methylation (RdDM) pathway (Espinas et al., 2016; Huang & Jin, 2022). These epigenetic marks may potentially be transmitted to subsequent generations via mechanisms involving small interfering RNAs (siRNAs) and chromatin remodelers, ensuring long-term stress resilience. However, the evidence regarding long-lasting transgenerational priming effects is still largely circumstantial (Harris et al.,

2023). Germline epigenetic plasticity, characterized by alterations in germline cells through modifications like DNA methylation and histone modifications, would further enhance evolvability and generate heritable phenotypic variation, a prerequisite for selection (Jablonka, 2013). This type of transgenerational epigenetic inheritance may potentially facilitate rapid and directional micro-evolutionary changes in plant populations, enabling them to adapt quickly to changing environmental conditions. Given its profound impact on plant resilience and potential for crop improvement, epigenetic stress memory holds immense promise for mitigating the detrimental effects of climate change and other environmental stressors. By elucidating the molecular mechanisms underlying the interplay between epigenetic pathways and germline epigenetic plasticity in mediating these responses, scientists can pave the way for developing original strategies to enhance crop yields and increase plant resistance to pests, diseases, and diverse environmental challenges.

Epigenetic Responses to Abiotic Stress

Drought

Drought stress can elicit dynamic alterations in the epigenetic landscape, modifying chromatin accessibility, DNA methylation and histone profiles, ultimately leading to transcriptional regulation (Forestan et al., 2018). In faba bean (*Vicia faba L.*), drought stress can induce changes in DNA methylation, leading to altered gene expression patterns in response to water deficit conditions (Abid et al., 2017). In *Arabidopsis*, stressful conditions triggers transient DNA hypermethylation of transposable elements near activated genes, defining the temporal sequence of transcriptional and epigenetic responses (Secco et al., 2015). These methylation changes may be transient or stable, depending on the specific target gene regions and the duration of the stress. The orchestrated epigenetic changes allow plants to fine-tune gene expression in response to drought stress, thereby enhancing their adaptability and survival. In apple, it has been shown that alterations in chromatin accessibility play a crucial role in the drought response by influencing the binding of RNA polymerase and regulatory factors (Wang et al., 2022). More than 200 apple genes were identified with enhanced chromatin accessibility and upregulated expression during drought stress, suggesting their potential roles in the drought response pathway (Wang et al., 2022). Plants may acquire a stress memory that translates into amplified transcriptional responses during subsequent drought encounters. These epigenetic responses can potentially affect the expression of stress-related genes, including those involved in water transport, stomatal regulation, and osmotic adjustment.

Histone modifications also participate in the response to drought stress. Changes in histone acetylation under water deficit conditions have been shown to affect the expression of stress-responsive genes (Li et al., 2021). Acetylation of core histones by histone acetyltransferases (HATs) loosens chromatin structure, typically promoting gene activation. Conversely, histone deacetylases (HDACs) remove acetyl groups, often leading to condensed chromatin and gene repression. For instance, overexpression of the histone deacetylase BdHD1 resulted in better survival under drought conditions in the model plant *Brachypodium distachyon* (Song et al., 2019). On the other hand, certain histone methylation marks like histone H3 lysine 4 trimethylation (H3K4me3) are associated with the activation of gene expression and influence the drought response. In barley, gene activation in response to drought stress involves the expression of genes related to the abscisic acid (ABA) pathway via H3K4me3 (Ost et al., 2023). Non-coding RNAs also play important roles in modulating drought stress responses. Specifically, small RNAs target genes involved in stress signaling, contributing to the optimization of plant responses to drought stress (Kumar et al., 2022; Guo et al., 2023). Unraveling the associations between epigenetic modifications and drought stress-related gene expression holds immense potential in crop improvement, especially in arid environments.

Temperature

Temperature stresses, including both extreme heat and cold, pose a serious threat to plant growth and survival. Epigenetic variation play a crucial role in plant responses to temperature, influencing thermotolerance and acclimation processes (Verma et al., 2022). Under heat stress conditions, plants may experience a variety of epigenetic changes that contribute to their response and adaptation. Specifically, it has been shown that the alteration of DNA methylation leads to changes in gene expression related to heat stress responses. Heat stress-induced DNA hypomethylation can be localized at target genomic regions, allowing for the upregulation of heat shock proteins (HSPs) and other stress-responsive genes (Ohama et al., 2017; Tiwari et al., 2023). Histone modifications like acetylation also contribute to the regulation of heat stress responses by orchestrating intricate epigenetic pathways. In *Arabidopsis*, the plant-specific histone deacetylases HD2B and HD2C have recently emerged as critical factors for heat stress-induced heterochromatin stabilization stress (Yang et al., 2023). They function through two distinct mechanisms: (1) promoting DNA methylation by associating with ARGONAUTE4 in nucleoli and Cajal bodies, and (2) facilitating nuclear accumulation of ARGONAUTE4. This dual role highlights both canonical (histone deacetylation) and non-canonical activities of HD2B and HD2C in maintaining heterochromatin integrity under heat stress (Yang et al., 2023).

Similar to high temperatures, cold stress can induce epigenetic modifications that enhance plant acclimation and tolerance to low temperatures. In rubber trees (*Hevea brasiliensis*), Tang et al. (2018) observed that cold treatment increases the transcriptional activity of demethylation-related genes *HbDME*, *HbROS*, and *HbDML*, influencing plant development and latex production. Moreover, rice exhibits genotype-specific cold-induced DNA methylation patterns that depend on growth stage and tissue, which may be mitotically heritable to contribute to cold stress responsiveness (Pan et al., 2011). Non-coding RNAs have also shown to play important roles in cold stress responses by modulating gene expression and facilitating the development of freezing tolerance. Kindgren et al. (2018) revealed that the lncRNA SVALKA regulates the expression of the C-repeat/dehydration-responsive element binding factor 1 (CBF 1) in *Arabidopsis*, ultimately enhancing its freezing tolerance without fitness costs. Similarly, miRNAs also appear to contribute to cold stress tolerance. Wang et al. (2014) demonstrated that overexpressing osa-miR319b enhances tolerance to cold stress in rice. Interestingly, the expression of potential osa-miR319b target genes, transcription factors OsPCF6 and OsTCP21, is significantly induced by cold stress. Remarkably, overexpression of osa-miR319b repressed this stress-induced upregulation. Furthermore, downregulation of OsPCF6 and OsTCP21, achieved through osa-miR319b overexpression, resulted in enhanced cold tolerance, partially via modulation of reactive oxygen species scavenging (Wang et al., 2014). Consequently, by developing crops that overexpress specific miRNAs associated with cold tolerance, breeders can potentially create varieties with improved performance under low-temperature conditions, ultimately increasing food security in areas vulnerable to cold snaps or frosts.

Salinity

High salinity is another major environmental stressor for plants, hindering growth and productivity. Epigenetic modifications are emerging as crucial players in plant adaptation upon such saline conditions. It has been shown that salinity stress can alter particular DNA methylation profiles, ultimately affecting the expression of genes involved in ion transport and osmotic stress responses (Skorupa et al., 2021). Wang et al. (2021) demonstrated that tetraploid rice exhibits enhanced salt tolerance due to epigenetic regulation of jasmonic acid (JA) pathway genes. Under salt stress, stress-responsive genes, including those in the JA pathway, are more rapidly induced and expressed at higher levels in tetraploid rice compared to diploid rice. Polyploidy can trigger DNA hypomethylation, potentially amplifying the expression of stress-responsive genes that often reside near TEs in the rice genome, whereas the elevated expression of stress-responsive genes in tetraploid rice can trigger hypermethylation following stress removal, leading to the suppression of transposable elements (TEs) located near these genes (Wang et al., 2021). Moreover, it is important to highlight that active DNA hypomethylation also plays a role in the response to salinity stress by inducing the expression of ion homeostasis-responsive genes. Li et al. (2023) identified *OsDML4*, a DNA demethylase gene, as essential for regulating rice salt tolerance. OsDML4 functions by regulating ROS homeostasis and the JA signaling pathway. This regulation is achieved through DNA methylation of the promoters of the key genes *OsCATA*, involved in ROS scavenging, and *OsJAZ13*, a component of the JA signaling pathway. They further revealed that CHH methylation within the *OsCATA* promoter region increased under salt stress. Remarkably, this increase was significantly higher in *osdml4* mutants compared to wild-type (Nipponbare) rice plants (Li et al., 2023).

Nutrient Deficits Stress

Nutrient imbalances, including both deficiencies and excesses, can disrupt normal epigenetic profiles, potentially leading to altered gene expression. Widiez et al. (2011) investigated this phenomenon in *Arabidopsis*. They focused on high nitrogen-insensitive 9-1 (*hni9-1*) mutant plants, which harbor a mutation in the *AtIWS1* gene. *AtIWS1* encodes a conserved component of the RNA polymerase II (RNAPII) elongation complex. The study revealed that *AtIWS1* acts in roots to repress the transcription of the gene *NRT2.1*, responsible for high-affinity nitrate uptake, under high nitrogen conditions. Furthermore, they observed that repression of *NRT2.1* by high nitrogen supply is associated with an increase in histone H3 lysine 27 trimethylation (H3K27me3) at the *NRT2.1 locus*, a repressive histone mark dependent on *AtIWS1*. Similarly, nutrient deficiencies, including phosphate (Pi) limitation, can also influence global DNA methylation patterns. Yong-Villalobos et al. (2015) observed extensive remodeling of DNA methylation in *Arabidopsis* under low Pi conditions. These changes often correlated with altered gene expression. Remarkably, modifications in the methylation pattern within gene regions frequently coincided with transcriptional activation or repression, highlighting the crucial role of dynamic methylation changes in regulating gene expression during Pi starvation (Yong-Villalobos et al., 2015). Phosphorus deficiency can also trigger changes in miRNA expression. Peng et al. (2020) observed that miR399d, the high-affinity Pi transporter McPHT1;4, and McMYB10 (a key transcription factor regulating leaf coloration) are strongly induced in leaves of *Malus* crabapple cultivars experiencing Pi deficiency.

Heavy Metal Contamination and Chemical Pollutants

Heavy metal contamination represents a significant environmental threat, impacting on plant growth and development. Recent research has highlighted the crucial role of epigenetic modifications in mediating plant responses to heavy metal toxicity, influencing organismal homeostasis, detoxification mechanisms, and stress tolerance (Cicatelli et al., 2014; Gallo-Franco et al., 2020; Fasani et al., 2023). DNA methylation exhibit dynamic alterations under contamination by heavy metals, contributing to changes in gene expression profiles associated with metal homeostasis and detoxification. Bölükbaşı and Karakaş (2023) investigated the effects of copper (Cu) exposure on DNA methylation patterns in safflower (*Carthamus tinctorius L.*) root tissues. Their findings revealed altered methylation patterns in response to varying Cu concentrations. These changes suggest a potential role for DNA methylation in safflower's tolerance to Cu toxicity. Feng et al. (2016) revealed that epigenetic variation in DNA methylation profiles is associated with gene expression in rice plants exposed to cadmium. Depending on the specific metal and genomic regions involved, heavy metals can induce either DNA hypomethylation or hypermethylation, ultimately influencing the expression of genes involved in metal transport, chelation, and stress responses. Jing et al. (2022) investigated DNA methylation responses in pokeweed, a model plant known to accumulate manganese and cadmium under heavy metal stress. Their findings revealed that different heavy metals target distinct DNA methylation sites in a dose-dependent manner. Furthermore, they demonstrated that reactive oxygen species (ROS) induced by heavy metal exposure partially mediate the changes in DNA methylation observed under manganese and cadmium stress (Jing et al., 2022).

Plant pollutants can trigger a cascade of detrimental effects, including physiological, morphological, and anatomical alterations. Air pollutants, particularly sulfur dioxide (SO2), ozone (O3), and nitrogen oxides (NOx) exert their negative impact on plant photosynthetic capacity by inducing stomatal closure, thereby restricting CO2 availability for photosynthesis (Baciak et al., 2015). Furthermore, reactive oxygen species (ROS) generated during pollutant-induced oxidative stress may also damage the photosynthetic apparatus. Pollutants enter plants through stomatal pores, leading to tissue injury and disruption of water balance (Robinson et al., 1998). These detrimental effects can trigger epigenetic alterations, ultimately impacting plant adaptation and detoxification processes. Epigenetic modifications, especially DNA methylation, hold promise as biomarkers for environmental pollution monitoring. Çiçekliyurt and Yayintas (2022) proposed DNA methylation levels in the bryophyte *Dicranum majus* Turner (greater fork-moss) as a potential biomarker for monitoring environmental pollution caused by heavy metals and toxic organic compounds. Similarly, Katsidi et al. (2023) employed an epigenetic analysis to investigate the effects of air pollution on black pine (*Pinus nigra*). Their study compared a control population to trees exposed to various pollutants, including SO2, NOx, and PM10 (particles with a diameter of 10 micrometres or less). Remarkably, the exposed population exhibited significantly higher DNA methylation levels in plant embryos compared to the control. Their findings suggest that air pollution may induce epigenetic alterations, potentially impacting its health and reproductive success.

Herbicides and insecticides (collectively categorized as pesticides) may also potentially affect the epigenetic machinery. Given their mutagenic effects, exposure to pesticides may lead to detrimental epigenetic alterations, affecting the expression of genes involved in oxidative stress, detoxification processes, and defense mechanisms. These epigenetic responses may serve as an adaptive strategy to mitigate the adverse effects of environmental stressors. In rice, Ma et al. (2021) revealed a role for the jasmonate signaling pathway in atrazine (ATZ) degradation. They demonstrated that CORONATINE INSENSITIVE 1a (OsCOI1a), a key component of this pathway, mediates ATZ degradation. ATZ exposure significantly induced OsCOI1a protein expression, ultimately enhancing tolerance to ATZ toxicity. Interestingly, the study also found that ATZ exposure triggered DNA hypomethylation and decreased histone H3K9me2 marks in the upstream region of the *OsCOI1a* gene. These epigenetic changes likely contribute to the enhanced expression of *OsCOI1a* under ATZ stress (Ma et al., 2021).

Epigenetic Responses to Biotic Stress

Pathogens

Plant-pathogen interactions involve a complex interplay between the plant immune system and pathogen virulence strategies. Epigenetic modifications can modulate pathogen recognition, defense signaling, and effector-triggered immunity. In *Arabidopsis*, Dowen et al. (2012) revealed that the virulent bacterial pathogen *Pseudomonas syringae* triggers DNA methylation changes across all target sequence contexts (CG, CHG, and CHH). Interestingly, this contrasted with the effects of an avirulent strain of the same bacteria, which only induced methylation changes in CG and CHG contexts (Dowen et al., 2012). Histone modifications also play a role in plant defense immunity responses. Crespo-Salvador et al. (2018) investigated the epigenetic response to the necrotrophic pathogen *Botrytis cinerea* in tomato plants. They found that the promoter regions and gene bodies of several defense-related genes, including *DES* (divinyl ethyl synthase), *LoxD* (lipoxygenase D), *DOX1* (α-dioxygenase 1), *PR2* (pathogenesis-related protein 2), *WRKY53*, and *WRKY33*, exhibited increased levels of the activating histone mark H3K4me3 upon infection with *B. cinerea* (Crespo-Salvador et al., 2018).

Plant adaptation involves complex interactions with herbivorous insects. Somers et al. (2023) demonstrated that cruciferous plants are armed with natural "epigenetic weapons" in the form of HDAC inhibitors like sulforaphane (SNF). This potent molecule disrupts the epigenetic machinery of herbivorous insects, offering a unique defense mechanism. The author observed that SNF suppresses HDAC enzymatic activity and slows the development of the pest *Spodoptera exigua*. Histone acetylation modifications have also been associated with the activation of defense genes, thereby enhancing the production of defensive compounds. Ding et al. (2012) demonstrated that manipulating the histone deacetylase HDT701 significantly alters plant defense responses in transgenic rice. Overexpression of HDT701 reduces H4 acetylation, making rice more susceptible to pathogens. Conversely, silencing HDT701 increases H4 acetylation, boosting pattern recognition receptor and defense gene expression, and ultimately strengthening rice against these biotic stress agents. Interestingly, Yu et al. (2013) proposed a novel role for DNA demethylation in plant immunity. Employing *Arabidopsis* as model, they observed that DNA demethylation might act as a priming mechanism to activate the transcription of defense genes potentially associated with TEs and repetitive sequences. Their findings revealed that during antibacterial defense, some TEs undergo demethylation and become transcriptionally reactivated. Consequently, DNA demethylation limited the spread of the bacterial pathogen *Pseudomonas syringae* within plant leaves. Furthermore, they observed that certain immune-response genes harboring repetitive sequences in their promoter regions were negatively regulated by DNA methylation (Yu et al., 2013).

Plants employ two key strategies to silence genes: transcriptional gene silencing (TGS) and post-transcriptional gene silencing (PTGS). TGS provides a permanent defense against DNA viruses through the RNA-directed DNA methylation (RdDM) pathway. In RdDM, non-coding RNA molecules guide DNA methylation, effectively silencing viral genes. PTGS, on the other hand, targets RNA molecules directly. These mechanisms function not only against RNA viruses but also in regulating the expression of the plant's own genes (Ashapkin et al., 2020). Plants can recognize double-stranded RNA (dsRNA) molecules, a hallmark of RNA virus infection, and ultimately trigger their degradation into small interfering RNAs (siRNAs) by DICER-like (DCL) family endoribonucleases (Ashapkin et al., 2020).

Epigenetic Reprogramming in Response to the Environment

Plants continuously adapt to fluctuating environmental conditions through a sophisticated array of adaptive strategies, encompassing dynamic epigenetic modifications, complex molecular crosstalk between genes, and hormonal signaling cascades. These intricate adaptive networks enable plants to adjust their physiological, metabolic, and developmental responses to external cues, including temperature fluctuations, light availability, nutrient gradients, and interactions with other organisms. If an environmental factor triggers an epigenetic change that is heritable through mitosis, it can have lasting phenotypic effects. This means that the altered gene expression pattern is copied and passed on to daughter cells, even if the original stressor disappears. Furthermore, mitotically heritable changes may potentially pass to germ line cells, being thereby transmitted to offspring through meiosis. This phenotype-environment interaction opens the door for environmental factors to influence not only the current generation but also future generations, even in the absence of the initial stressor (Turner, 2009).

Epigenetic Mechanisms Mediating Plant Adaptation: Flowering and Vernalization Response via Flowering Locus C (FLC)

Floral transition is environmentally controlled by epigenetic-driven pathways involving Flowering Locus C (FLC), a transcription factor that represses floral regulatory genes. Prolonged cold exposure progressively silences the *FLC* gene via epigenetic modifications, and this silencing can be stably maintained (Baulcombe & Dean, 2014). Elucidating *FLC* regulation in the *Arabidopsis* life cycle has yielded insights broadly applicable to plant floral transition. It has also been proposed that this knowledge can greatly contribute to developing crops resilient to climate change and maintaining yield stability (Nielsen, 2021). In winter *Arabidopsis* accessions, the late flowering and vernalization requirement arises from high *FLC* expression mediated by activation via FRIGIDA (FRI) (Michaels & Amasino, 1999). FRIGIDA (FRI) is a well-characterized protein that promotes high transcriptional activity of

FLC during early plant development by interacting with chromatin modifiers and RNA maturation factors, thereby preventing the premature termination of *FLC* transcription (Choi et al., 2011; Geraldo et al., 2009; Schon et al., 2021). The winter annual *Arabidopsis* accessions harbor an active *FRI* allele, leading to elevated *FLC* expression. Facilitated by SUF4 binding to the *FLC* promoter, FRI is recruited to the *locus*, forming the FRI complex (Choi et al., 2011). SUF4 directly targets the *FLC* promoter via *cis*-element recognition. FRI then acts as a scaffold, assembling FRL1, FES1, and FLX proteins into a functional transcriptional activator complex (Choi et al., 2011). This mechanism is crucial as FRI antagonizes the activity of the FLC repressor, FCA, very early during embryo development (Schon et al., 2021). FCA is a RNA-binding protein that interacts with the SWI/SNF chromatin remodeler SWI3B. Macknight et al. (1997) revealed that AtSWI3B interacts with FCA through the C-terminal region of FCA, located outside its RNA-binding domain. Remarkably, FCA is a critical component of the flowering pathway that acts as an inductor of flowering by preventing the accumulation of *FLC* transcripts (Quesada et al., 2005). FCA promotes proximal *FLC* transcript termination, maintaining a repressive chromatin state and low *FLC* expression (Schon et al., 2021). In contrast, the presence of FRI disrupts FCA activity, resulting in an open chromatin landscape and high *FLC* expression, thereby establishing the vernalization requirement for flowering (Schon et al., 2021).

Crevillén et al. (2013) investigated the involvement of higher-order chromatin structures in FLC regulation. They identified a gene loop formed by the physical interaction of the *FLC locus*'s 5′ and 3′ flanking regions. Interestingly, Polycomb-mediated epigenetic silencing induced by vernalization coincided with the disruption of the loop. This disruption mirrored the timing of the cold-induced *FLC* transcriptional shutdown and upregulation of the *FLC* antisense transcript *COOLAIR*, an lncRNA transcribed from the *FLC locus* and proposed to promote *FLC* silencing (Trotman et al., 2021). Consequently, this molecular pathway exemplifies how an environmental signal (vernalization) facilitates the silencing of a key gene (*FLC*) through the generation of antisense lncRNA transcribed from its own *locus*. While we already characterized such an intricate interplay shaping the plant's flowering response, our understanding of gene regulation continues evolving, with evidence suggesting this type of regulation may be more widespread than previously recognized. Moreover, it has been shown that C-repeat (CRT)/dehydration-responsive elements (DREs) at the 3′-end of *FLC* are required for cold-mediated expression of *COOLAIR* (Jeon et al., 2023). CRT/DRE elements, characterized by a conserved 5-bp core sequence (CCGAC), are frequently found in the promoter regions of *Arabidopsis* genes responsive to cold and dehydration stress. Remarkably, these elements are also present in the promoters of *COR* (cold-regulated) genes (Jaglo et al., 2001).

Epigenetic reprograming during FLC reactivation primarily depends on cold exposure, which triggers rapid down-regulation of *FLC* transcription, a process mediated by *COOLAIR*. This is followed by a switch to epigenetic silencing involving the conserved Polycomb repressive complex 2 (PRC2), and its accessory proteins. Remarkably, it has been shown that PRC2-mediated silencing occurs independently of DNA methylation (Costa & Dean, 2019). PRC2 deposits the repressive histone mark H3K27me3 on target genes, leading to transcriptional repression. During DNA replication, Polycomb-mediated silencing spreads across the *FLC locus*, locking in the silenced state through multiple cell divisions. This stable silencing can be reversed through histone demethylation (Costa & Dean, 2019). Crevillén et al. (2014) revealed that EARLY FLOWERING 6 (ELF6) is involved in epigenetic reprograming during FLC reactivation. ELF6 is an evolutionally conserved demethylase associated with decreased di- and tri-methylated H3K27 levels. They identified a mutation in ELF6 that disrupts FLC reactivation in reproductive tissues and leads to partial inheritance of a vernalized state (Crevillén et al., 2014). ELF6 mutants exhibited reduced H3K27me3 demethylase activity and consequently the next generation of mutant plants showed higher H3K27me3 levels and lower *FLC* expression compared to wild-type plants. Reduced *FLC* expression facilitates a rapid-cycling reproductive strategy, which is potentially advantageous in suitable climates. Beyond flowering regulation, other proteins that interact with *FLC* play vital roles in plant development. In 2024, Huang et al. observed that SUF4, a C2H2 zinc finger protein which represses flowering via *FLC*, also interacts with EBS, a bivalent histone reader, to shape root development. SUF4 and EBS are recruited to *SCARECROW* (*SCR*), a key root development regulator. They promote H3K4me3 (activating) and suppress H3K27me3 (repressing) marks on *SCR*, leading to its activation and influencing quiescent center specification in the root apical meristem (Huang et al., 2024).

Environmental Challenges and Epigenetic Variation

Environmental stimuli trigger specific epigenetic modifications mediated by DNA methylation, histone modifications, and targeted ncRNA-related pathways, ultimately leading to precise fine-tuning of gene expression upon specific conditions. As above described, these changes may potentially be heritable across generations. Deciphering the intricate interplay between specific environmental stimuli and plant responses at the level of epigenetic marks is crucial for comprehending the molecular mechanisms underpinning plant adaptation and resilience (Rapp & Wendel, 2005). This deeper understanding, in turn, holds immense potential for developing novel genetic tools for crop improvement, including the development of stress-resistant crop varieties and optimization of agricultural yields and quality under changing environmental conditions.

Exploring the Potential for Transgenerational Inheritance

A diverse set of environmental stressors, including drought, extreme temperatures, pollutants, and pathogen attacks, can trigger dynamic alterations in plant epigenetic landscapes. Notably, these epigenetic marks may potentially be inherited across generations through both sexual reproduction, via germinal cells, and vegetative propagation, via structures like stolons and tubers. Transgenerational epigenetic inheritance allows higher organisms to "remember" past environmental challenges and pre-adapt their progeny for analogous conditions. This inheritance mechanism paves the way for rapid and efficient adaptation when recurrent environmental stressors arise (Boyko & Kovalchuk, 2011). The inheritance of specific epigenetic marks associated with stress responses can provide the descendants with enhanced resilience and adaptability, allowing them to confront analogous challenges with greater efficiency. Transgenerational epigenetic inheritance offers a still unexplored layer of plasticity and adaptability to plant evolution. While genetic modifications unfold over extended periods through the process of natural selection, epigenetic modifications could facilitate swift adaptation to rapidly changing environments. However, not all environmentally induced epigenetic changes exhibit the stability required for reliable inheritance. The persistence of epigenetic modifications, both within and across generations, depends on several key factors, including the degree and predictability of environmental fluctuations, dispersal patterns, and the epigenetic landscape governing phenotypic responses to environmental cues (Herman et al., 2014).

The molecular mechanisms underpinning transgenerational epigenetic inheritance are a subject of ongoing investigation (Fitz-James & Cavalli, 2022). This intriguing process likely involves the establishment and robust maintenance of specific epigenetic marks within germ cells. These epigenetic modifications may need then be faithfully recognized and preserved throughout embryogenesis and subsequent development by distinct epigenetic mechanisms. DNA methylation in plants is a multifaceted process regulated by RNA-directed mechanisms. These pathways recruit DNA methylation machinery to precise genomic locations, resulting in *locus*-specific methylation patterns. The DNA methylation maintenance system ensures the high-fidelity inheritance of methylated cytosines onto newly synthesized strands during DNA replication, while the RdDM pathway introduces new epigenetic marks through targeted methylation of unmethylated cytosines guided by small RNAs (Tirot & Jullien, 2022). The ability to tailor stress responses across generations, informed by the epigenetic memory of past exposures (e.g., drought, salinity, heat stress), represents the cornerstone that defines the phenomenon of transgenerational adaptive plasticity (Boyko & Kovalchuk, 2011). In response to evolutionary trajectories, plants may potentially exhibit differential levels of transgenerational epigenetic inheritance for a given trait, thereby evidencing variation in the efficiency and fidelity of epigenetic mark transmission across generations.

Epigenetics Mechanisms and Environmental Conditions

Plants possess a remarkable ability to sense and respond to environmental cues. This sensitivity translates into phenotypic changes mediated by a complex interplay of cellular receptors, signal transduction pathways, and downstream effector molecules. Environmental stresses further trigger dynamic epigenetic reprogramming. Plants orchestrate this response through specific DNA methylation patterns, histone post-translational modifications (PTMs), such as the activating H3K4me3 and repressive H3K27me3 marks, and ncRNA-mediated pathways. Epigenetic adjustments can lead to rapid and precise changes in gene expression patterns, enabling plants to dynamically adapt to diverse environmental conditions. Sun et al. (2021) demonstrated a direct link between DNA methylation and dehydration stress memory, shedding light on the molecular mechanisms underlying drought acclimation-induced tolerance in *Boea hygrometrica*, a key model organism for understanding plant responses to this environmental condition (dehydration). Moreover, stress-responsive genes involved in salt tolerance may be activated through specific histone acetylation, enhancing the plant's ability to withstand adverse conditions and maintain its survival. Luo et al. (2012) identified HD2C and HDA6 as key players in plant salt tolerance, demonstrating their interaction to suppress ABA-responsive genes and thereby modulate stress resilience. Plants hold epigenetic variants associated with fine-tuned gene expression profiles and orchestrated stress responses, ultimately enhancing their adaptability and survival under fluctuating environmental conditions. By identifying these variants and unlocking the molecular mechanisms underlying the interplay between epigenetic pathways, we can develop crops adapted to drought, saline stress, and other environmental adversities. This not only will safeguard food security for future generations but also pave the way for a more sustainable and climate-resilient agricultural future.

Conclusions

To effectively address the challenges posed by rapid climate change, comprehending the intricate relationship between epigenetic variation, stress memory, and plant development is imperative. Epigenetic mechanisms offer plants a remarkable degree of flexibility, allowing them to adapt to environmental pressures. By remembering past stresses through epigenetic modifications, plants can prime themselves for future challenges, potentially increasing resilience across generations. This knowledge presents exciting possibilities for manipulating epigenetic variation to develop crops better equipped to handle the adversities of climate change.

References

Abid, G., Mingeot, D., Muhovski, Y., Mergeai, G., Aouida, M., Abdelkarim, S., et al. (2017). Analysis of DNA methylation patterns associated with drought stress response in faba bean (Vicia faba L.) using methylation-sensitive amplification polymorphism (MSAP). *Environmental and Experimental Botany, 142,* 34–44.

Annacondia, M. L., & Martinez, G. (2019). Plant models of transgenerational epigenetic inheritance. In *Transgenerational epigenetics* (pp. 263–282). Academic.

Ashapkin, V. V., Kutueva, L. I., Aleksandrushkina, N. I., & Vanyushin, B. F. (2020). Epigenetic mechanisms of plant adaptation to biotic and abiotic stresses. *International Journal of Molecular Sciences, 21*(20), 7457.

Baciak, M., Warmiński, K., & Bęś, A. (2015). The effect of selected gaseous air pollutants on woody plants. *Leśne Prace Badawcze.*/Forest research papers December 2015, *76*(4), 401–409.

Baduel, P., & Sasaki, E. (2023). The genetic basis of epigenetic variation and its consequences for adaptation. *Current Opinion in Plant Biology, 75,* 102409.

Banerjee, A., & Roychoudhury, A. (2017). Epigenetic regulation during salinity and drought stress in plants: Histone modifications and DNA methylation. *Plant Gene, 11,* 199–204.

Baulcombe, D. C., & Dean, C. (2014). Epigenetic regulation in plant responses to the environment. *Cold Spring Harbor Perspectives in Biology, 6*(9), a019471.

Bölükbaşı, E., & Karakaş, M. (2023). Modeling DNA methylation profiles and epigenetic analysis of safflower (Carthamus tinctorius L.) seedlings exposed to copper heavy metal. *Toxics, 11*(3), 255.

Boyko, A., & Kovalchuk, I. (2011). Genome instability and epigenetic modification—Heritable responses to environmental stress? *Current Opinion in Plant Biology, 14*(3), 260–266.

Cao, S., Wang, L., Han, T., Ye, W., Liu, Y., Sun, Y., et al. (2022). Small RNAs mediate transgenerational inheritance of genome-wide trans-acting epialleles in maize. *Genome Biology, 23*(1), 53.

Choi, K., Kim, J., Hwang, H. J., Kim, S., Park, C., Kim, S. Y., & Lee, I. (2011). The FRIGIDA complex activates transcription of FLC, a strong flowering repressor in Arabidopsis, by recruiting chromatin modification factors. *The Plant Cell, 23*(1), 289–303.

Cicatelli, A., Todeschini, V., Lingua, G., Biondi, S., Torrigiani, P., & Castiglione, S. (2014). Epigenetic control of heavy metal stress response in mycorrhizal versus non-mycorrhizal poplar plants. *Environmental Science and Pollution Research, 21,* 1723–1737.

Çiçekliyurt, M. M. H., & Yayintas, O. T. (2022). DNA methylation in bryophytes as a biomarker for monitoring environmental pollution. *Indian Journal of Experimental Biology (IJEB), 60*(11), 870–874.

Costa, S., & Dean, C. (2019). Storing memories: The distinct phases of Polycomb-mediated silencing of Arabidopsis FLC. *Biochemical Society Transactions, 47*(4), 1187–1196.

Crespo-Salvador, Ó., Escamilla-Aguilar, M., López-Cruz, J., López-Rodas, G., & González-Bosch, C. (2018). Determination of histone epigenetic marks in Arabidopsis and tomato genes in the early response to Botrytis cinerea. *Plant Cell Reports, 37,* 153–166.

Crevillén, P., Sonmez, C., Wu, Z., & Dean, C. (2013). A gene loop containing the floral repressor FLC is disrupted in the early phase of vernalization. *The EMBO Journal, 32*(1), 140–148.

Crevillén, P., Yang, H., Cui, X., Greeff, C., Trick, M., Qiu, Q., et al. (2014). Epigenetic reprogramming that prevents transgenerational inheritance of the vernalized state. *Nature, 515*(7528), 587–590.

Ding, B., Bellizzi, M. D. R., Ning, Y., Meyers, B. C., & Wang, G. L. (2012). HDT701, a histone H4 deacetylase, negatively regulates plant innate immunity by modulating histone H4 acetylation of defense-related genes in rice. *The Plant Cell, 24*(9), 3783–3794.

Dowen, R. H., Pelizzola, M., Schmitz, R. J., Lister, R., Dowen, J. M., Nery, J. R., et al. (2012). Widespread dynamic DNA methylation in response to biotic stress. *Proceedings of the National Academy of Sciences, 109*(32), E2183–E2191.

Dubin, M. J., Zhang, P., Meng, D., Remigereau, M. S., Osborne, E. J., Paolo Casale, F., et al. (2015). DNA methylation in Arabidopsis has a genetic basis and shows evidence of local adaptation. *eLife, 4*, e05255.

Espinas, N. A., Saze, H., & Saijo, Y. (2016). Epigenetic control of defense signaling and priming in plants. *Frontiers in Plant Science, 7*, 1201.

Fasani, E., Giannelli, G., Varotto, S., Visioli, G., Bellin, D., Furini, A., & DalCorso, G. (2023). Epigenetic control of plant response to heavy metals. *Plants, 12*(18), 3195.

Feng, S. J., Liu, X. S., Tao, H., Tan, S. K., Chu, S. S., Oono, Y., et al. (2016). Variation of DNA methylation patterns associated with gene expression in rice (Oryza sativa) exposed to cadmium. *Plant, Cell & Environment, 39*(12), 2629–2649.

Fitz-James, M. H., & Cavalli, G. (2022). Molecular mechanisms of transgenerational epigenetic inheritance. *Nature Reviews Genetics, 23*(6), 325–341.

Forestan, C., Farinati, S., Lunardon, A., & Varotto, S. (2018). Integrating transcriptome and chromatin landscapes for deciphering the epigenetic regulation of drought response in maize. In J. Bennetzen, S. Flint-Garcia, C. Hirsch, & R. Tuberosa (Eds.), *The Maize Genome. Compendium of plant genomes* (pp. 97–112). Springer.

Gallo-Franco, J. J., Sosa, C. C., Ghneim-Herrera, T., & Quimbaya, M. (2020). Epigenetic control of plant response to heavy metal stress: A new view on aluminum tolerance. *Frontiers in Plant Science, 11*, 602625.

Geraldo, N., Bäurle, I., Kidou, S. I., Hu, X., & Dean, C. (2009). FRIGIDA delays flowering in Arabidopsis via a cotranscriptional mechanism involving direct interaction with the nuclear cap-binding complex. *Plant Physiology, 150*(3), 1611–1618.

Godwin, J., & Farrona, S. (2020). Plant epigenetic stress memory induced by drought: A physiological and molecular perspective. *Methods in Molecular Biology, 2093*, 243–259.

Guo, S., Xu, T., Ju, Y., Lei, Y., Zhang, F., Fang, Y., et al. (2023). MicroRNAs behave differently to drought stress in drought-tolerant and drought-sensitive grape genotypes. *Environmental and Experimental Botany, 207*, 105223.

Harris, C. J., Amtmann, A., & Ton, J. (2023). Epigenetic processes in plant stress priming: Open questions and new approaches. *Current Opinion in Plant Biology, 75*, 102432.

Herman, J. J., Spencer, H. G., Donohue, K., & Sultan, S. E. (2014). How stable 'should' epigenetic modifications be? Insights from adaptive plasticity and bet hedging. *Evolution, 68*(3), 632–643.

Hollister, J. D., & Gaut, B. S. (2009). Epigenetic silencing of transposable elements: A trade-off between reduced transposition and deleterious effects on neighboring gene expression. *Genome Research, 19*(8), 1419–1428.

Huang, C. Y., & Jin, H. (2022). Coordinated epigenetic regulation in plants: A potent managerial tool to conquer biotic stress. *Frontiers in Plant Science, 12*, 795274.

Huang, C., Wang, D., Yang, Y., Yang, H., Zhang, B., Li, H., et al. (2024). SUPPRESSOR OF FRIGIDA 4 cooperates with the histone methylation reader EBS to positively regulate root development. *Plant Physiology, kiae321*.

Jablonka, E. (2013). Epigenetic inheritance and plasticity: The responsive germline. *Progress in Biophysics and Molecular Biology, 111*(2–3), 99–107.

Jaglo, K. R., Kleff, S., Amundsen, K. L., Zhang, X., Haake, V., Zhang, J. Z., et al. (2001). Components of the Arabidopsis C-repeat/dehydration-responsive element binding factor cold-response pathway are conserved in Brassica napus and other plant species. *Plant Physiology, 127*(3), 910–917.

Jeon, M., Jeong, G., Yang, Y., Luo, X., Jeong, D., Kyung, J., et al. (2023). Vernalization-triggered expression of the antisense transcript COOLAIR is mediated by CBF genes. *eLife, 12*, e84594.

Jing, M., Zhang, H., Wei, M., Tang, Y., Xia, Y., Chen, Y., et al. (2022). Reactive oxygen species partly mediate DNA methylation in responses to different heavy metals in pokeweed. *Frontiers in Plant Science, 13*, 845108.

Katsidi, E. C., Avramidou, E. V., Ganopoulos, I., Barbas, E., Doulis, A., Triantafyllou, A., & Aravanopoulos, F. A. (2023). Genetics and epigenetics of Pinus nigra populations with differential exposure to air pollution. *Frontiers in Plant Science, 14*, 1139331.

Kindgren, P., Ard, R., Ivanov, M., & Marquardt, S. (2018). Transcriptional read-through of the long non-coding RNA SVALKA governs plant cold acclimation. *Nature Communications, 9*(1), 4561.

Kumar, S., Das, M., Sadhukhan, A., & Sahoo, L. (2022). Identification of differentially expressed mungbean miRNAs and their targets in response to drought stress by small RNA deep sequencing. *Current Plant Biology, 30*, 100246.

Li, S., He, X., Gao, Y., Zhou, C., Chiang, V. L., & Li, W. (2021). Histone acetylation changes in plant response to drought stress. *Genes, 12*(9), 1409.

Li, C., Kong, J. R., Yu, J., He, Y. Q., Yang, Z. K., Zhuang, J. J., et al. (2023). DNA demethylase gene OsDML4 controls salt tolerance by regulating the ROS homeostasis and the JA signaling in rice. *Environmental and Experimental Botany, 209*, 105276.

Luo, M., Wang, Y. Y., Liu, X., Yang, S., Lu, Q., Cui, Y., & Wu, K. (2012). HD2C interacts with HDA6 and is involved in ABA and salt stress response in Arabidopsis. *Journal of Experimental Botany, 63*(8), 3297–3306.

Ma, L. Y., Zhai, X. Y., Qiao, Y. X., Zhang, A. P., Zhang, N., Liu, J., & Yang, H. (2021). Identification of a novel function of a component in the jasmonate signaling pathway for intensive pesticide degradation in rice and environment through an epigenetic mechanism. *Environmental Pollution, 268*, 115802.

Macknight, R., Bancroft, I., Page, T., Lister, C., Schmidt, R., Love, K., et al. (1997). FCA, a gene controlling flowering time in Arabidopsis, encodes a protein containing RNA-binding domains. *Cell, 89*(5), 737–745.

Michaels, S. D., & Amasino, R. M. (1999). FLOWERING LOCUS C encodes a novel MADS domain protein that acts as a repressor of flowering. *The Plant Cell, 11*(5), 949–956.

Nielsen, M. (2021). *Deciphering the role of PRC2 accessory proteins in promoting cold-induced epigenetic switching in Arabidopsis thaliana.* (Doctoral dissertation,. University of East Anglia.

Ohama, N., Sato, H., Shinozaki, K., & Yamaguchi-Shinozaki, K. (2017). Transcriptional regulatory network of plant heat stress response. *Trends in Plant Science, 22*(1), 53–65.

Ost, C., Cao, H. X., Nguyen, T. L., Himmelbach, A., Mascher, M., Stein, N., & Humbeck, K. (2023). Drought-stress-related reprogramming of gene expression in barley involves differential histone modifications at ABA-related genes. *International Journal of Molecular Sciences, 24*(15), 12065.

Pan, Y., Wang, W., Zhao, X., Zhu, L., Fu, B., & Li, Z. (2011). DNA methylation alterations of rice in response to cold stress. *Plant Omics Journal, 4*(7), 364–369.

Peng, Z., Tian, J., Luo, R., Kang, Y., Lu, Y., Hu, Y., et al. (2020). MiR399d and epigenetic modification comodulate anthocyanin accumulation in Malus leaves suffering from phosphorus deficiency. *Plant, Cell & Environment, 43*(5), 1148–1159.

Quadrana, L., & Colot, V. (2016). Plant transgenerational epigenetics. *Annual Review of Genetics, 50*, 467–491.

Quadrana, L., Bortolini Silveira, A., Mayhew, G. F., LeBlanc, C., Martienssen, R. A., Jeddeloh, J. A., & Colot, V. (2016). The Arabidopsis thaliana mobilome and its impact at the species level. *eLife, 5*, e15716.

Quesada, V., Dean, C., & Simpson, G. G. (2005). Regulated RNA processing in the control of Arabidopsis flowering. *The International Journal of Developmental Biology, 49*(5–6), 773–780.

Rapp, R. A., & Wendel, J. F. (2005). Epigenetics and plant evolution. *New Phytologist, 168*(1), 81–91.

Robinson, M. F., Heath, J., & Mansfield, T. A. (1998). Disturbances in stomatal behaviour caused by air pollutants. *Journal of Experimental Botany, 49*, 461–469.

Schon, M., Baxter, C., Xu, C., Enugutti, B., Nodine, M. D., & Dean, C. (2021). Antagonistic activities of cotranscriptional regulators within an early developmental window set FLC expression level. *Proceedings of the National Academy of Sciences, 118*(17), e2102753118.

Secco, D., Wang, C., Shou, H., Schultz, M. D., Chiarenza, S., Nussaume, L., et al. (2015). Stress induced gene expression drives transient DNA methylation changes at adjacent repetitive elements. *eLife, 4*, e09343.

Skorupa, M., Szczepanek, J., Mazur, J., Domagalski, K., Tretyn, A., & Tyburski, J. (2021). Salt stress and salt shock differently affect DNA methylation in salt-responsive genes in sugar beet and its wild, halophytic ancestor. *PLoS One, 16*(5), e0251675.

Somers, D. J., Kushner, D. B., McKinnis, A. R., Mehmedovic, D., Flame, R. S., & Arnold, T. M. (2023). Epigenetic weapons in plant-herbivore interactions: Sulforaphane disrupts histone deacetylases, gene expression, and larval development in Spodoptera exigua while the specialist feeder Trichoplusia ni is largely resistant to these effects. *PLoS One, 18*(10), e0293075.

Song, J., Henry, H. A., & Tian, L. (2019). Brachypodium histone deacetylase BdHD1 positively regulates ABA and drought stress responses. *Plant Science, 283*, 355–365.

Sun, R. Z., Liu, J., Wang, Y. Y., & Deng, X. (2021). DNA methylation-mediated modulation of rapid desiccation tolerance acquisition and dehydration stress memory in the resurrection plant Boea hygrometrica. *PLoS Genetics, 17*(4), e1009549.

Tang, X., Wang, Q., Yuan, H., & Huang, X. (2018). Chilling-induced DNA demethylation is associated with the cold tolerance of Hevea brasiliensis. *BMC Plant Biology, 18*, 1–16.

Tirot, L., & Jullien, P. E. (2022). Epigenetic dynamics during sexual reproduction: At the nexus of developmental control and genomic integrity. *Current Opinion in Plant Biology, 69*, 102278.

Tiwari, A., Pandey-Rai, S., Rai, K. K., Tiwari, A., & Pandey, N. (2023). Molecular and epigenetic basis of heat stress responses and acclimatization in plants. *The Nucleus, 66*(1), 69–79.

Trotman, J. B., Braceros, K. C., Cherney, R. E., Murvin, M. M., & Calabrese, J. M. (2021). The control of polycomb repressive complexes by long noncoding RNAs. *Wiley Interdisciplinary Reviews: RNA, 12*(6), e1657.

Turgut-Kara, N., Arikan, B., & Celik, H. (2020). Epigenetic memory and priming in plants. *Genetica, 148*, 47–54.

Turner, B. M. (2009). Epigenetic responses to environmental change and their evolutionary implications. *Philosophical Transactions of the Royal Society B: Biological Sciences, 364*(1534), 3403–3418.

Vaschetto, L. M. (2016). Miniature Inverted-repeat Transposable Elements (MITEs) and their effects on the regulation of major genes in cereal grass genomes. *Molecular Breeding, 36*(3), 30.

Verma, N., Giri, S. K., Singh, G., Gill, R., & Kumar, A. (2022). Epigenetic regulation of heat and cold stress responses in crop plants. *Plant Gene, 29*, 100351.

Wang, S. T., Sun, X. L., Hoshino, Y., Yu, Y., Jia, B., Sun, Z. W., et al. (2014). MicroRNA319 positively regulates cold tolerance by targeting OsPCF6 and OsTCP21 in rice (Oryza sativa L.). *PLoS One, 9*(3), e91357.

Wang, L., Cao, S., Wang, P., Lu, K., Song, Q., Zhao, F. J., & Chen, Z. J. (2021). DNA hypomethylation in tetraploid rice potentiates stress-responsive gene expression for salt tolerance. *Proceedings of the National Academy of Sciences, 118*(13), e2023981118.

Wang, S., He, J., Deng, M., Wang, C., Wang, R., Yan, J., et al. (2022). Integrating ATAC-seq and RNA-seq reveals the dynamics of chromatin accessibility and gene expression in apple response to drought. *International Journal of Molecular Sciences, 23*(19), 11191.

Weinhold, A. (2018). Transgenerational stress-adaption: An opportunity for ecological epigenetics. *Plant Cell Reports, 37*, 3–9. https://doi.org/10.1007/s00299-017-2216-y

Widiez, T., El Kafafi, E. S., Girin, T., Berr, A., Ruffel, S., Krouk, G., et al. (2011). HIGH NITROGEN INSENSITIVE 9 (HNI9)-mediated systemic repression of root NO3− uptake is associated with changes in histone methylation. *Proceedings of the National Academy of Sciences, 108*(32), 13329–13334.

Yang, F., Sun, Y., Du, X., Chu, Z., Zhong, X., & Chen, X. (2023). Plant-specific histone deacetylases associate with ARGONAUTE4 to promote heterochromatin stabilization and plant heat tolerance. *New Phytologist, 238*(1), 252–269.

Yong-Villalobos, L., González-Morales, S. I., Wrobel, K., Gutiérrez-Alanis, D., Cervantes-Peréz, S. A., Hayano-Kanashiro, C., et al. (2015). Methylome analysis reveals an important role for epigenetic changes in the regulation of the Arabidopsis response to phosphate starvation. *Proceedings of the National Academy of Sciences, 112*(52), E7293–E7302.

Yu, A., Lepère, G., Jay, F., Wang, J., Bapaume, L., Wang, Y., et al. (2013). Dynamics and biological relevance of DNA demethylation in Arabidopsis antibacterial defense. *Proceedings of the National Academy of Sciences, 110*(6), 2389–2394.

Yu, Y., Zhang, Y., Chen, X., & Chen, Y. (2019). Plant noncoding RNAs: Hidden players in development and stress responses. *Annual Review of Cell and Developmental Biology, 35*, 407–431.

Zhao, J., Lu, Z., Wang, L., & Jin, B. (2020). Plant responses to heat stress: Physiology, transcription, noncoding RNAs, and epigenetics. *International Journal of Molecular Sciences, 22*(1), 117.

Chapter 4
Heritable Epigenetic Phenomena

Plants exhibit fascinating heritable epigenetic phenomena that transcend the boundaries of Mendelian inheritance. Epigenetic phenomena, which include paramutation, imprinting, and transgenerational epigenetic inheritance, allow plants to adapt and fine-tune gene expression across generations. They offer exciting possibilities for understanding and potentially manipulating heritable traits under challenging conditions. Paramutation could be harnessed to induce stable silencing of negative alleles, while imprinting could guide breeding strategies to optimize gene expression from parental lines. Moreover, transgenerational epigenetic inheritance holds promise for developing crops pre-adapted to future environmental challenges. By unraveling the molecular mechanisms underlying such phenomena, we can unlock novel strategies to improve crop resilience and adaptation, ultimately enhancing food security upon variable conditions.

Paramutation

Paramutation can be defined as a complex epigenetic interaction between two alleles at a single *locus*. This epigenetic phenomenon involves the heritable silencing of one allele by another, achieved through the transfer of repressive marks, leaving the DNA sequence unaltered. As a result, the silenced allele's altered expression can propagate within the population, contradicting the Mendelian principles of genetics. The allele inducing these changes is called the paramutagenic allele, while the susceptible one that undergoes modification is termed the paramutable allele.

Paramutation has been observed in diverse organisms, including animals and plants like maize and tomato, where it has been shown to affect various traits, including pigmentation and flowering. In plants, small RNAs guide the RdDM pathway to target specific genomic regions for methylation, leading to heritable paramutation. These regulatory RNAs often originate from one allele capable of initiating gene silencing *in trans* through mechanisms involving histone modifications (Stam, 2009; Arteaga-Vazquez & Chandler, 2010). Tandem repeat units within the genome can act as the target in the paramutable (silenced) allele (Stam et al., 2002). This targeted silencing, in turn, leads to the establishment of a stable, heritable paramutagenic state.

L. M. Vaschetto, *Epigenetics in Crop Improvement*, https://doi.org/10.1007/978-3-031-73176-1_4

Epigenetic modifications, particularly DNA methylation and small RNA interactions, orchestrate the transmission of paramutagenic and paramutable alleles. The target *loci* structure, including palindromes and methylation within coding regions, modulates its susceptibility to these mechanisms, facilitating the formation of meiotically transmissible epialleles (Brzeski & Brzeska, 2011; Khaitová et al., 2011). In plants, it has been shown that not only DNA methylation, but also nucleosome occupancy and histone modifications, are involved in the establishment and maintenance of the silenced state at the paramutable *loci* (Haring et al., 2010; Pilu, 2015). Moreover, chromatin-remodeling factors have also been involved in paramutation. In maize, Deans et al. (2020) identified the CHD3 protein, orthologous to the *Arabidopsis* chromatin-remodeling factor PICKLE, as essential for maintaining repression of the paramutable *purple plant1* (*pl1*) allele.

By elucidating the intricate interplay between DNA methylation, small RNA silencing, chromatin modifications, and chromatin remodeler components, we not only expand our understanding of the epigenetic mechanisms involved in paramutation but also unlock its potential for engineering gene silencing-mediated states associated with desired crop traits. This knowledge holds promise to revolutionize agricultural practices by enabling the development of crops with enhanced resistance to disease, pests, and environmental stressors, leading to increased yields and enhanced nutritional value in a sustainable manner.

The study of paramutation in plants has also provided invaluable insights into the regulation and activity of regulatory non-coding RNAs in transgenerational epigenetics, expanding our understanding of epigenetic inheritance and the establishment of heritable epigenetic states. This phenomenon involves dynamic reprogramming of the epigenome at paramutable *loci*, encompassing alterations in DNA methylation patterns mediated by small RNAs and the RdDM pathway. The deposition of specific methylation marks guided by small regulatory RNAs acts as a molecular "memory" that account for the transgenerational persistence of the silenced state and its associated phenotypic consequences (Erhard Jr & Hollick, 2011). This epigenetic memory is further reinforced by histone modifications, which potentially contribute to chromatin remodeling and the stabilization of the silenced state (Haring et al., 2010).

Plant genomes encode different classes of non-coding RNAs involved in gene silencing pathways, including those involved in paramutation. The abundance and diversity of non-coding RNAs vary among plant species, suggesting coevolution between environmental adaptations and gene-silencing mechanisms. The maize RNA polymerase IV (RNAP IV) is well-known to safeguard repressive heterochromatin chromatin states via small interfering RNAs (siRNAs) that target repetitive sequences like TEs, triggering *de novo* cytosine methylation and histone modifications (Deans et al., 2020). RNAP IV has been implicated in paramutation of *b1, r1, p1,* and *pl1* genes, and both mediators of paramutation (*mop*) of the *B1*-Intense (*B1-I*) and factors required to maintain repression (*rmr*) of the *Pl1-Rhoades* alleles, which encode for MOP/RMR proteins mediating or maintaining paramutation, are either RNAP IV subunits or 24-nucleotide (24 nt) RNA biogenesis factors (Deans et al., 2020). These observations provide evidence that small RNAs act as primary messengers for the transference of epigenetic information between alleles involved in paramutation.

Unlocking Paramutation in Crops: Maize as Study Case

Maize provided the initial stage for unraveling the epigenetic mechanisms involved in paramutation. A canonical example of paramutation in maize is the booster 1 (*b1*) *locus*. The *b1* gene encodes a transcription factor essential for activating the purple anthocyanin biosynthesis pathway. Paramutation at the *b1 locus* involves paramutable *B-I* and paramutagenic *B'* (epi)alleles, which are associated with intense and reduced plant pigmentation, respectively (Aubert et al., 2022). Figure 4.1 illustrates the correlation between such a paramutation event and its associated phenotypic alterations regarding tissue pigmentation. Paramutation at the maize *b1 locus* involves a cross between the low-expressing paramutagenic *B'* and high-expressing paramutable *B-I* epialleles, which results in complete and stable silencing of the *B-I* epiallele in the F1 generation. The resulting paramutated *B'** allele (*B-I* in the previous generation) acquires paramutagenic capacity, leading to exclusively light-pigmented progeny when crossed with a paramutable genotype (Hövel et al., 2015).

Despite identical nucleotide sequences, the *B-I* and *B'* alleles exhibit distinct DNA methylation patterns, with the *B-I* allele characterized by a more open chromatin conformation compared to *B'* (Stam et al., 2002). The *b1 locus* contains seven unique noncoding transcribed tandem repeats of 853 bp located approximately 100 Kb upstream of the transcription start, which are required for paramutation-mediated silencing of the *B'* epiallele in a process involving the production of small interfering RNAs (siRNAs) (Belele et al., 2013). Both siRNAs and protein components of the RdDM pathway (e.g., DRM2, Pol VI, Pol V) are required to mediate paramutation, evidencing the role of different factors in this epigenetic phenomenon. The evolutionary conserved ARGONAUTE (AGO) proteins identified as members of the RdDM effector complex are involved in paramutation at the *b1 locus* (Belele et al., 2013; Giacopelli & Hollick, 2015). Specifically, AGO104 binds to 24-nucleotide small interfering RNAs (24-nt siRNAs) involved in RdDM (Aubert et al., 2022). The establishment of paramutation at the *b1 locus* requires several RdDM pathway factors, namely the RNA-dependent RNA polymerase MOP1, the NRP(D/E)2a subunit encoded by *mop2*, and the NRPD1 subunit encoded by *mop3* (Hövel et al., 2024). Bidirectional transcripts originating from tandem repeats undergo DCL and MOP1-mediated processing to generate siRNAs, which subsequently induce paramutation (Chandler, 2007).

Histone modifications dictate chromatin landscapes and gene accessibility, ultimately influencing paramutation outcomes. In the case of the paramutable *b1 locus*, it is marked with repressive DNA methylation marks, facilitating heterochromatin formation (Haring et al., 2010). The paramutagenic allele orchestrates the deposition of these marks on the paramutable *locus*, leading to gene silencing and the establishment of the paramutated state. The transgenerational inheritance of

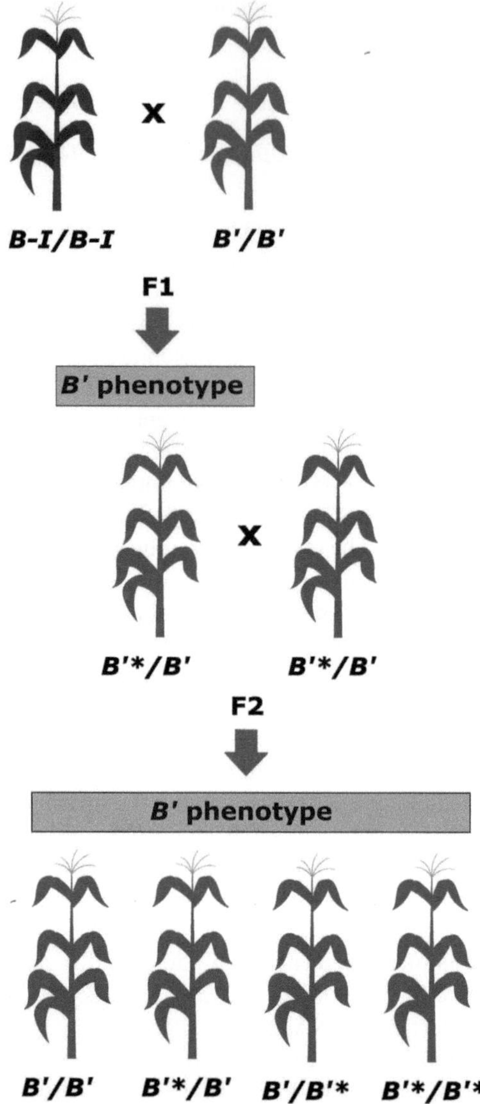

Fig. 4.1 Effects on the phenotype (tissue color) linked to paramutation at the *b1 locus* of maize (*Zea mays*). In maize, the *b1* gene functions as a transcriptional regulator, controlling the anthocyanin pigment pathway. The level of *b1* gene activity directly determines how much purple pigment a plant produces. Specific DNA sequences within the *b1 locus* control its expression through paramutation (Stam et al., 2002). When a highly pigmented plant (homozygous for the *B-I* allele) interacts with a less pigmented version (homozygous for *B'*), the highly pigmented one always gets turned into a less pigmented plant (*B'**) (Coe, 1959). Thus, *B-I* is "paramutated" by *B'*, which is now denoted as *B'**. According to a Mendelian point of view, crosses between heterozygous (*B'*/B'*) plants in the F2 would be expected to produce 25% pigmented plants; however, they only produce lightly pigmented plants. The new *B'** (epi)allele has the same ability to induce paramutation as the original *B'* allele

epigenetic modifications ensures the stability of the epigenetic memory. Additionally, nucleosome occupancy, H3 acetylation, and methylation at H3K9 and H3K27 are mainly involved in tissue-specific regulation of the tandem (hepta) repeat sequence, ultimately also modulating epigenetic regulation at the *b1 locus* (Haring et al., 2010).

Paramutation in Molecular Breeding

The *trans*-inactivation cascade orchestrated by paramutation, mediated primarily by non-coding RNAs and epigenetic modifications at target *loci*, raises a powerful tool for molecular breeding. This phenomenon transcends the limitations of genetic modification, potentially facilitating the improvement of specific plant traits by inducing persistent modifications in gene expression patterns without altering the underlying DNA sequence. The paramutagenic allele's ability to catalyze the deposition of repressive epigenetic marks at the paramutable *locus* triggers gene silencing and establishes a paramutated state that persists transgenerationally. This capacity of paramutation to induce transgenerational phenotypic alterations, independent of Mendelian inheritance, highlights its potential for crop improvement. The stability of the paramutable allele warrants careful evaluation on a case-by-case basis when translating these results into practical crop applications.

Genomic Imprinting

Genomic imprinting is an fascinating epigenetic phenomenon that establishes parent-specific gene expression patterns through epigenetic modifications. It was first discovered in flowering plants and has now been extensively identified and studied in organisms of different taxonomic groups, including therian mammals and some arthropods, ultimately revealing an evolutionary pattern across different kingdoms. Genomic imprinting may have arisen from selection pressures favoring parental manipulation of resource allocation (Costa et al., 2012). While the precise mechanisms governing imprinting in plants remain elusive (McKeown et al., 2014), parental epigenetic marks, primarily differential methylation, are crucial for establishing and maintaining genomic imprinting.

Genomic imprinting entails the inheritance of allele-specific epigenetic modifications for specific chromosomal segments, originating from either the paternal or maternal germline. It arises primarily from distinct epigenetic marks established in male and female gametes before fertilization. The endosperm, which mediates the relationship between the maternal parent and the embryo, is the primary site of gene imprinting in flowering plants. Pre-determined epigenetic modifications on parental alleles within the endosperm dictate monoallelic expression, with the DNA-methylated paternal allele often being transcriptionally inactive, while the demethylated maternal allele becomes transcriptionally active (Lu et al., 2022). DNA methylation serves as a recruitment signal for chromatin-modifying complexes, leading to the establishment of repressive histone modifications like H3K27 methylation, which further suppress gene expression (Iwasaki et al., 2019). These epigenetic modifications can alter chromatin structure and accessibility, contributing to the establishment and perpetuation of imprinting marks. Given its importance during seed development, understanding the interplay between imprinted *loci* and epigenetic modifications can potentially pave the way for epigenetic-based approaches

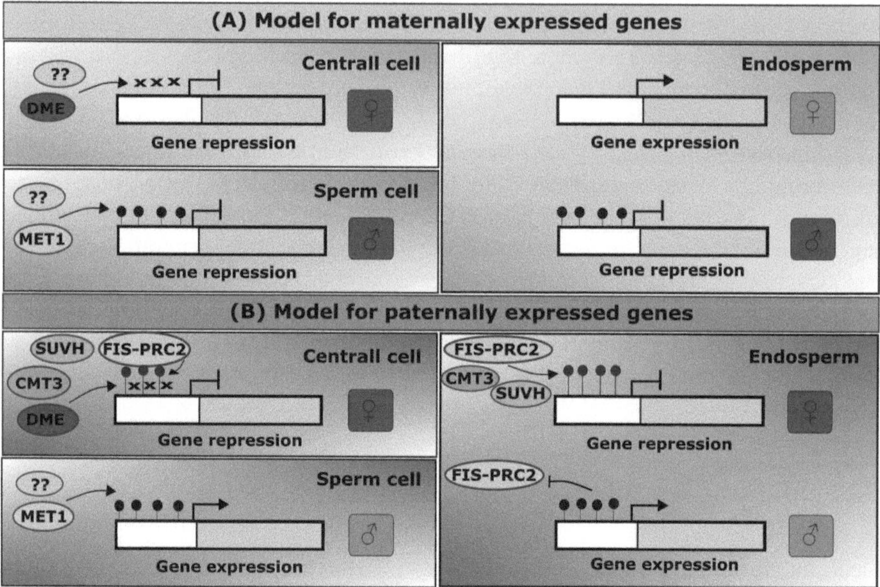

Fig. 4.2 Models of imprinted gene regulation at maternally and paternally expressed *loci* (adapted from Batista & Köhler, 2020). (**a**) Model for maternally expressed genes in the endosperm. These maternally expressed genes exhibit expression only in the endosperm, while their transcription is repressed in sporophytic tissues. Constitutive DNA methylation marks on maternally expressed genes enforce their silencing in sporophytic tissues. To facilitate maternal expression in the endosperm, maternal DNA methylation is removed, and paternal methylation is maintained through the concerted actions of DME and MET1, respectively. This model offers a simplified view, regardless the potential influence of other epigenetic mechanisms such as the RdDM pathway. (**b**) Model for paternally expressed genes (expression both in endosperm and sporophytic tissues). According to this model, paternally expressed genes carry constitutive DNA methylation marks. In this case, DNA methylation does not induce transcriptional silencing but instead prevents H3K27me3 repressive marks by the FIS-PRC2 complex. Maternal-specific DNA demethylation, mediated by DME, enables subsequent H3K27me3 deposition on these alleles, resulting in their transcriptional repression within the endosperm. Conversely, the presence of DNA methylation on paternally expressed genes precludes H3K27me3 deposition, allowing for gene expression. Plant SU(VAR)3–9 homologs (SUVHs) are histone methyltransferases that mediate repressive methylation on lysine 9 of histone H3 (H3K9) at heterochromatin. CHROMOMETHYLASE3 (CMT3), a plant-specific DNA methyltransferase, is responsible for catalyzing DNA methylation within the CHG sequence context. In the diagram, X marks depict the removal of DNA methylation marks. DNA methylation marks: shadow circles, histone repressive marks: blue circles

to improve crop yields. Figure 4.2 presents recently revisited models for the regulation of maternally and paternally expressed *loci* involved in imprinting (adapted from Batista & Köhler, 2020).

Unraveling imprinting in flowering plants necessitates a comprehensive understanding of the process of double fertilization. This process, involving identical sperm cells and distinct female gametes (egg and central cell), results in the formation of the triploid endosperm (two maternal and one paternal genome) and the diploid embryo. The embryo transmits genetic information while the endosperm nourishes the embryo for development and germination. Genomic imprinting, mainly documented in the endosperm and with few reported cases in the embryo, may involve DNA hypomethylation mediated by DEMETER (DME) and histone modifications mediated by Polycomb group (PcG) proteins. The maternal central cell expresses the DNA glycosylase DEMETER (DME) to actively remove imprints at specific *loci* before fertilization, targeting methylated cytosines regardless of the sequence context (Batista & Köhler, 2020). Post-fertilization, the endosperm inherits the hypomethylated central cell genome, establishing asymmetric DNA methylation between maternal and paternal contributions (Choi et al., 2002; Iwasaki et al., 2022). Consequently, parent-of-origin-specific expression has been linked to differential DNA methylation states in both endosperm and embryos (Jahnke & Scholten, 2009), although the mechanisms underlying the establishment of imprinting in plant embryos remain less understood.

Genomic imprinting hinges on the complex interplay between DNA methylation and distinct histone modifications (Satyaki & Gehring, 2017; Batista & Köhler, 2020). Moreno-Romero et al. (2019) revealed that paternally expressed genes in the *Arabidopsis* endosperm exhibit a unique epigenetic signature characterized by H3K27me3 deposition mediated by Polycomb Repressive Complex 2 (PRC2), in conjunction with heterochromatic H3K9me2 and CHG methylation, specifically targeting silenced maternal alleles of these genes. In contrast to *Arabidopsis*, maternal alleles of paternally expressed genes in maize are marked by H3K27me3 but lack CHG methylation (Zhang et al., 2014). The maternal small RNAs regulate spatial and temporally imprinting profiles linked to seed development. In *Arabidopsis*, Kirkbride et al. (2019) identified maternal-specific siRNAs capable of fine-tuning endosperm development by silencing transcription factors AGAMOUS-LIKE (AGL). The authors linked these siRNA-targeted gene expression patterns to seed size and imprinting, ultimately evidencing potential for crop improvement through manipulation of the siRNA-mediated RdDM pathway. Small RNAs derived from Transposable Elements and related repeat sequences can direct DNA methylation and repressive histone modifications in a sequence-specific manner, leading to the repression of certain genomic regions (Kirkbride et al., 2019).

Imprinting in Arabidopsis

Genomic imprinting exhibits a fascinating diversity of mechanisms, encompassing bipartite methylation, antisense transcription, and parent-of-origin-specific chromatin modifications (details on these molecular mechanisms are beyond the scope of this book). As a model organism, *Arabidopsis* provides a unique opportunity to investigate the evolutionary dynamics and functional implications of imprinting in plants. Here, I centered on the largely known imprinted genes *MEA* and *PHE1*.

Maternal Control of Embryogenesis (MEA, also known as MEDEA), a maternally expressed but paternally imprinted gene, exemplifies the interplay between epigenetic mechanisms and their functional roles in seed development. This gene represses embryo and endosperm growth in the absence of fertilization. MEA is a member of the Polycomb Repressive Complex 2 (PRC2), which exhibits histone methyltransferase activity by methylating histone H3 on lysine 27 residues (H3K27me3). MEA harbors a SET domain with methyltransferase activity that mediates stable transcriptional silencing of target genes (Grossniklaus et al., 1998). Endosperm imprinting dictates parent-specific expression of the Polycomb gene *MEA*, with the maternal allele actively transcribed and the paternal allele silenced. The DNA glycosylase DME activates *MEA*, while the DNA methyltransferase MET1 maintains CG methylation at the *MEA locus*, regulating its imprinting (Gehring et al., 2006). Through a feedback loop mechanism, MEA autoregulates its imprinted expression, allele-specifically and dynamically during seed development (Gehring et al., 2006, TAIR, 2023). Importantly, the *MEA* gene has also been associated with stress response. Roy et al. (2018) observed that, upon pathogen challenge, its transcription is induced, leading to the limitation of an excessive immune response. Consequently, MEA acts as a feedback inhibitor of defense, contributing to the plant's growth-defense balance.

Beyond histone marks, additional post-translational modifications contribute to the regulation of parental imprinting. Dumbliauskas et al. (2011) revealed that the *Arabidopsis* CUL4-DDB1 complex, known for its role in protein ubiquitylation, interacts with MSI1 and is essential for maintaining *MEA* imprinting. Köhler et al. (2003) identified MSI1 as a component of the FIS-Polycomb repressive complex 2 (PRC2) complex, alongside MEDEA (MEA), FERTILIZATION-INDEPENDENT SEED2 (FIS2), and FERTILIZATION-INDEPENDENT ENDOSPERM (FIE). The FIS-PRC2 complex exerts multiple roles associated with cell identity specification and developmental stage transitions, ultimately contributing to mitotic repression and endosperm cellularization (Hands et al., 2016). Intriguingly, *cul4* mutant plants display autonomous endosperm initiation and a loss of parental imprinting for *MEA*. This finding evidences a functional connection between CUL4-DDB1 and PRC2 complexes, ultimately suggesting the role of distinct posttranslational modifications in the establishment of gene expression patterns during imprinting.

The imprinted gene *PHERES1* (*PHE1*) have expanded the complexity of parental-specific epigenetic regulation mechanisms in *Arabidopsis*. This MADS-box transcription gene, required for transcriptional activation of key genes in the endosperm, is called *'Pheres'* in memory of one of the murdered sons of the mythological 'Medea' (Köhler et al., 2003). The maternal *PHE1* allele likely serves as a recruitment site for the MEA-containing FIS-PRC2 complex in the central cell before fertilization (Köhler et al., 2005). This complex operates in both the female gametophyte and endosperm to prevent premature *PHE1* activation before fertilization and to confine *PHE1* expression to the endosperm's chalazal region post-fertilization. This repressive function is correlated to H3K27me3 deposition at the *PHE1 locus* (Köhler et al., 2003; Makarevich et al., 2006, 2008). Makarevich et al. (2008) proposed that differential DNA methylation of the *PHE1* 3′ region, with maternal allele hypomethylation in the central cell, regulates imprinting at this *locus*. In this regard, it has already been shown that a 54 bp triple tandem repeat within the 3′ region of the *PHE1 locus* is required for control of its imprinting (Villar et al., 2009). In the endosperm, the unmethylated 3′ region, in conjunction with associated protein complexes, may inhibit transcription through chromatin looping or silencer recruitment. Conversely, in pollen, methylation of this region prevents repressive chromatin formation or silencer function (Makarevich et al., 2008). The direct repeats within the *PHE1* 3′ region exhibit preferential CG methylation, thereby suggesting a dependency on the MET1 methyltransferase for maintaining this methylation pattern (Makarevich et al., 2008).

Imprinting in Maize

The first imprinted gene discovered was the *red-color* (*R*) gene in maize endosperm (Kermicle, 1970). This gene, currently referred to as *r1*, exhibits maternal expression, as evidenced by the production of fully colored kernels when specific alleles are inherited maternally (Selinger & Chandler, 2001; Gehring, 2013). The number of imprinted genes identified in maize has expanded to hundreds since this initial discovery (Batista & Köhler, 2020). The extensive repertoire of imprinted genes in maize underscores their roles in regulating seed development, holding the potential for developing seeds with improved traits through targeted manipulation of imprinting mechanisms. While predominantly characterized in the endosperm, genomic imprinting has also been documented in maize embryos. By employing high-throughput RNA sequencing, Meng et al. (2018) analyzed gene expression patterns in maize embryos derived from reciprocal crosses between inbred lines B73 and Mo17. They identified 64 imprinted genes, including 57 and 7 maternally and paternally expressed genes, respectively, with a high number of them expressed in the early stages of embryo development.

Imprinted genes represent putative targets for manipulating seed traits to improve crop yield, quality, and stress tolerance. Dai et al. (2022) identified *ded1*, an imprinted *locus* in maize, as a paternal regulator of seed size and development. Interestingly, hypomorphic *ded1* alleles, when transmitted through the male parent, result in a 5–10% seed weight reduction, underscoring the gene's dosage-dependent effect. DED1 directly promotes the expression of essential genes for early endosperm development, ultimately influencing grain size (Dai et al., 2022).

Imprinting and Seed Development

In the endosperm, imprinted gene expression is primarily observed in genes derived primarily from one parental genome. In the embryo, gene imprinting has been observed at various stages of development, with a greater number of imprinted genes identified at early embryo stages. Imprinted genes play a pivotal role in orchestrating seed size, exerting precise control over the molecular mechanisms governing endosperm proliferation and cellularization. Even weak alterations in their expression patterns can elicit significant changes in seed size, impacting crop yield potential. Yuan et al. (2017) identified over 250 parentally-differentially expressed genes in the rice endosperm, and over 100 of them linked to grain yield quantitative trait *loci* (QTLs). Remarkably, these imprinted genes have been associated with miniature inverted-repeat transposable elements (MITEs), differentially methylated regions (DMRs), and a subset of siRNAs and lncRNAs. Furthermore, imprinted genes may also exhibit a critical role in directing nutrient storage within seeds. These genes may be associated with the accumulation of vital compounds such as starch, proteins, oils, and other metabolites, suggesting that imprinting holds promise for enhancing crop nutritional quality. Xin et al. (2013) revealed a critical role for maternally expressed genes in maize endosperm at 10 days post-pollination, coinciding with the initiation of starch and storage protein accumulation. These genes likely facilitate nutrient uptake and allocation during this crucial stage of seed development.

By fine-tuning the expression of genes involved in the biosynthesis of nutrients, imprinted genes can potentially enrich the seeds with valuable health-promoting compounds. This opens up exciting possibilities for developing crops with enhanced nutritional value, contributing to improved food security and dietary health. Imprinted genes noy only play a crucial role in seed size and shape the nutritional profile of seeds, but also influence characteristics like dormancy and germination (Piskurewicz et al., 2016). By regulating the expression of genes involved in hormonal signaling and stress responses, imprinted genes can also determine how long a seed remains dormant before germinating and how it responds to environmental cues. Hence, precise epigenetic manipulation of imprinted gene expression can help in the design of crops with optimized seed size, nutrient quality, germination characteristics, and broader adaptability.

Transgenerational Epigenetic Inheritance

Transgenerational epigenetic inheritance plays a pivotal role in plant biology and offers profound insights into adaptation, evolution, and the transmission of epigenetic information across generations. This phenomenon allows plants to transmit acquired information from previous generations to their offspring, providing a mechanism for rapid adaptation to changing environmental conditions. The flexibility provided by transgenerational epigenetic inheritance mechanisms may potentially allow plants to respond to environmental cues and optimize their characteristics for survival and reproductive success. In *Arabidopsis*, *FWA* (*FLOWERING WAGENINGEN*), a floral repressor, is silenced in the wild type (WT) by promoter hypermethylation, leading to early flowering (Soppe et al., 2000). Conversely, *fwa-1* mutants lacking methylation exhibit late flowering. Interestingly, FWA expression is restricted to the endosperm in WT plants, remaining silent in somatic tissues (Kinoshita et al., 2004). Kinoshita et al. (2007) observed that mutations in the chromatin remodeler *DDM1* (*decrease in DNA methylation 1*) lead to ectopic *FWA* expression and imprinting disruption. DDM1 is a member of the SWI2/SNF2 chromatin remodeling family, known to be associated with histone modifications and DNA methylation. Remarkably, transcribing double-stranded RNA targeting the *FWA* promoter triggered *de novo* methylation, transcriptional silencing, and phenotypic reversion. This heritability correlated with the inheritance of CG methylation patterns in the promoter, hinting at a potential mechanism for maintaining epigenetic memory beyond the initial inducing signal (Kinoshita et al., 2007).

Phenotypic Plasticity and Transgenerational Epigenetic Inheritance

As mentioned before in this book, phenotypic plasticity, the capacity of plants to adjust their traits in response to environmental signals, is closely associated with epigenetic pathways. Once viewed as a source of variation, plasticity is now recognized as a critical adaptation. Both mechanisms share potential underlying regulatory pathways influencing gene expression. Environmental heterogeneity and generational stability emerge as key factors shaping the evolution of both epigenetic inheritance and phenotypic plasticity (Lind & Spagopoulou, 2018). Epigenetic modifications can regulate gene expression patterns that influence plastic responses, allowing plants to adapt their growth, development, and physiological processes to environmental conditions. This plasticity enhances their ability to thrive in various habitats under changing climates and contributes to the survival and persistence of plant populations in challenging environments. Transgenerational epigenetic inheritance also holds significant implications for plant evolution. Epigenetic changes accumulated over generations can result in stable variations in physical traits, influencing the direction and speed of evolutionary processes. By transmitting epigenetic information across generations, plants can expedite the adaptation process to new environments and contribute to the generation of diversity.

Heritable differences in epigenetic regulation can be triggered by both external (environmental) and internal (genomic) stresses and signals. In particular, internally induced variations often exhibit stability, offering a potential avenue for crop improvement (Cao & Chen, 2024). Recognizing the relevance of transgenerational epigenetic inheritance in plant adaptation is of fundamental importance in crop improvement. By unraveling the mechanisms of epigenetic inheritance, we can unlock molecular breeding strategies and biotechnological tools to enhance crop yield and resilience in the face of environmental adversity (Cao & Chen, 2024). This knowledge will pave the way to develop innovative approaches for enhancing crop performance, bolstering stress tolerance, and optimizing resource allocation.

Epigenetic modifications plays a crucial role in transmitting information associated with optimal developmental strategies across generations. In plants, transgenerational epigenetic inheritance exert significant effects on developmental transitions, for example, during flowering, seed germination, and root architecture formation. These transgenerational effects may also amplify phenotypic plasticity by conveying information on environmental cues and optimal responses to subsequent generations. This would enable offspring to respond more effectively to specific environmental conditions. Consequently, stress-induced epigenetic modifications passed down through generations, can serve as a record of past environmental challenges. Additionally, transgenerational memory would enhance the phenotypic response of offspring, enabling them to better withstand similar stresses. Transgenerational epigenetic effects may thus contribute to adaptive plasticity by providing a mechanism for swift adaptation to changing environments. By inheriting epigenetic modifications linked to specific environmental cues, offspring potentially can exhibit traits better suited to prevailing conditions, facilitating their survival, reproduction, and persistence in diverse habitats.

Epigenetic Reprogramming and Transgenerational Inheritance

Land plants exhibit a unique life cycle, alternating between diploid and haploid generations. The transition from diploid to haploid is marked by the loss of H3K9me2 histone marks and DNA demethylation of *cis*-regulatory elements near transposable elements (TEs) (Borg et al., 2021). These changes trigger dramatic shifts in chromatin accessibility and transcriptional landscapes. The haploid phase features a global loss of H3K27me3, establishing a chromatin state primed for the return to diploidy after fertilization (Borg et al., 2021). TE silencing is crucial for maintaining genome integrity across generations, and plants achieve this through heritable DNA methylation patterns (Law & Jacobsen, 2010). Remarkably, plants exhibit meiotic inheritance of gene silencing, suggesting a transgenerational silencing mechanism. Moreover, epigenetic reprogramming in non-germline reproductive cells may potentially reinforce TE silencing in germ cells (Feng et al., 2010). These observations highlight the multifaceted strategies employed by plants to ensure TE control and genome stability.

Epigenetic reprogramming is essential not only for silencing TEs but also for other critical functions (e.g., ensuring meiosis competence, establishing genomic imprinting, etc.). In plants, epigenetic reprogramming in the germ line is a distinct process compared to mammals. In mammals, reprogramming occurs in the zygote and germ line during early development. Conversely, plants do not produce germ cells until later in development, thereby epigenetic reprogramming occurs via different mechanisms. Recent advances in genome-wide epigenetic analyses have provided insight into the dynamic regulation of epigenetic states during plant reproductive development. Epigenetic regulation plays a crucial role in the formation and function of plant gametes, including the establishment of unique epigenetic landscapes in different gametes.

During epigenetic reprogramming, DNA methylation marks and histone modifications are reinstated in a coordinated fashion. This reprogramming process facilitates the establishment of lineage-specific epigenetic patterns and the transmission of transgenerational memory to future generations. Disruptions in the epigenetic reprogramming process potentially can bear significant consequences for further development. Errors or incomplete reprogramming may lead to the inheritance of inappropriate or altered epigenetic marks, resulting in phenotypic anomalies or compromised responses to environmental cues in offspring. Such disruptions should have negative impacts on plant development, adaptation, and overall fitness.

Applications and Implications in Crop Improvement

Transgenerational epigenetic inheritance harbors immense potential crop improvement, including but not limited to enhanced agronomic yield, stress resilience, and disease resistance. By inducing and manipulating transgenerational memory, breeders potentially can develop crops that are better adapted to specific environmental conditions, ultimately increasing their resilience and productivity. Similarly, the transmission of stress-responsive epigenetic marks to offspring would enable plants to be primed for improved adaptability to adverse conditions such as drought, heat, or salinity. Furthermore, comprehending the transgenerational inheritance of disease resistance can inform the development of strategies for disease management. By inducing transgenerational memory, plants can convey augmented resistance traits to their progeny, contributing to durable disease control and reducing reliance on pesticides. Elucidating the specific epigenetic signatures linked to desirable phenotypic outcomes and dissecting their transmission dynamics represents a crucial step toward harnessing the potential of epigenetic transgenerational inheritance phenomena.

Epigenetic reprogramming during early plant development also offers a window for intervention. Molecular breeders can potentially introduce targeted molecules or environmental cues to establish favorable epigenetic marks that enhance a crop's resilience or productivity. Hence, by understanding how epigenetic reprogramming influences traits like stress tolerance or flowering time, we can develop methods to induce or manipulate these epigenetic changes, leading to improved yields and adaptation.

Conclusion

Paramutation, imprinting, and transgenerational epigenetic inheritance collectively represent a promising frontier in crop improvement. By understanding and manipulating such epigenetic phenomena we can develop crop varieties with enhanced yields, superior nutritional quality and higher resilience to environmental stressors, including those associated with climate change. Therefore, they represent *per se* epigenetic tools with potential to rapidly adapt crops to changing conditions without relying solely on traditional molecular breeding methods. Consequently, the ability to stably inherit phenotypic modifications through their epigenetic pathways provides new avenues for crop optimization, ultimately contributing to global food security.

References

Arteaga-Vazquez, M. A., & Chandler, V. L. (2010). Paramutation in maize: RNA mediated trans-generational gene silencing. *Current Opinion in Genetics & Development, 20*(2), 156–163.

Aubert, J., Bellegarde, F., Oltehua-Lopez, O., Leblanc, O., Arteaga-Vazquez, M. A., Martienssen, R. A., & Grimanelli, D. (2022). AGO104 is a RdDM effector of paramutation at the maize b1 locus. *PLoS One, 17*(8), e0273695.

Batista, R. A., & Köhler, C. (2020). Genomic imprinting in plants—Revisiting existing models. *Genes & Development, 34*(1–2), 24–36.

Belele, C. L., Sidorenko, L., Stam, M., Bader, R., Arteaga-Vazquez, M. A., & Chandler, V. L. (2013). Specific tandem repeats are sufficient for paramutation-induced trans-generational silencing. *PLoS Genetics, 9*(10), e1003773.

Borg, M., Papareddy, R. K., Dombey, R., Axelsson, E., Nodine, M. D., Twell, D., & Berger, F. (2021). Epigenetic reprogramming rewires transcription during the alternation of generations in Arabidopsis. *eLife, 10*, e61894.

Brzeski, J., & Brzeska, K. (2011). The maze of paramutation: A rough guide to the puzzling epigenetics of paramutation. *Wiley Interdisciplinary Reviews: RNA, 2*(6), 863–874.

Cao, S., & Chen, Z. J. (2024). Transgenerational epigenetic inheritance during plant evolution and breeding. *Trends in Plant Science.*

Chandler, V. L. (2007). Paramutation: From maize to mice. *Cell, 128*(4), 641–645.

Choi, Y., Gehring, M., Johnson, L., Hannon, M., Harada, J. J., Goldberg, R. B., et al. (2002). DEMETER, a DNA glycosylase domain protein, is required for endosperm gene imprinting and seed viability in Arabidopsis. *Cell, 110*(1), 33–42.

Coe Jr, E. H. (1959). A regular and continuing conversion-type phenomenon at the B locus in maize. *Proceedings of the National Academy of Sciences, 45*(6), 828–832.

Costa, L. M., Yuan, J., Rouster, J., Paul, W., Dickinson, H., & Gutierrez-Marcos, J. F. (2012). Maternal control of nutrient allocation in plant seeds by genomic imprinting. *Current Biology, 22*(2), 160–165.

Dai, D., Mudunkothge, J. S., Galli, M., Char, S. N., Davenport, R., Zhou, X., et al. (2022). Paternal imprinting of dosage-effect defective1 contributes to seed weight xenia in maize. *Nature Communications, 13*(1), 5366.

Deans, N. C., Giacopelli, B. J., & Hollick, J. B. (2020). Locus-specific paramutation in Zea mays is maintained by a PICKLE-like chromodomain helicase DNA-binding 3 protein controlling development and male gametophyte function. *PLoS Genetics, 16*(12), e1009243.

Dumbliauskas, E., Lechner, E., Jaciubek, M., Berr, A., Pazhouhandeh, M., Alioua, M., et al. (2011). The Arabidopsis CUL4–DDB1 complex interacts with MSI1 and is required to maintain MEDEA parental imprinting. *The EMBO Journal, 30*(4), 731–743.

Erhard, K. F., Jr., & Hollick, J. B. (2011). Paramutation: A process for acquiring trans-generational regulatory states. *Current Opinion in Plant Biology, 14*(2), 210–216.

Feng, S., Jacobsen, S. E., & Reik, W. (2010). Epigenetic reprogramming in plant and animal development. *Science, 330*(6004), 622–627.

Gehring, M. (2013). Genomic imprinting: Insights from plants. *Annual Review of Genetics, 47*, 187–208.

Gehring, M., Huh, J. H., Hsieh, T. F., Penterman, J., Choi, Y., Harada, J. J., et al. (2006). DEMETER DNA glycosylase establishes MEDEA polycomb gene self-imprinting by allele-specific demethylation. *Cell, 124*(3), 495–506.

Giacopelli, B. J., & Hollick, J. B. (2015). Trans-homolog interactions facilitating paramutation in maize. *Plant Physiology, 168*(4), 1226–1236.

Grossniklaus, U., Vielle-Calzada, J. P., Hoeppner, M. A., & Gagliano, W. B. (1998). Maternal control of embryogenesis by MEDEA, a polycomb group gene in Arabidopsis. *Science, 280*(5362), 446–450.

Hands, P., Rabiger, D. S., & Koltunow, A. (2016). Mechanisms of endosperm initiation. *Plant Reprod, 29*, 215–225. https://doi.org/10.1007/s00497-016-0290-x

Haring, M., Bader, R., Louwers, M., Schwabe, A., van Driel, R., & Stam, M. (2010). The role of DNA methylation, nucleosome occupancy and histone modifications in paramutation. *The Plant Journal, 63*(3), 366–378.

Hövel, I., Pearson, N. A., & Stam, M. (2015, August). Cis-acting determinants of paramutation. In *Seminars in cell & developmental biology* (Vol. 44, pp. 22–32). Academic.

Hövel, I., Bader, R., Louwers, M., Haring, M., Peek, K., Gent, J. I., & Stam, M. (2024). RNA-directed DNA methylation mutants reduce histone methylation at the paramutated maize booster1 enhancer. *Plant Physiology, 195*(2), 1161–1179.

Iwasaki, M., Hyvärinen, L., Piskurewicz, U., & Lopez-Molina, L. (2019). Non-canonical RNA-directed DNA methylation participates in maternal and environmental control of seed dormancy. *eLife, 8*, e37434.

Iwasaki, M., Penfield, S., & Lopez-Molina, L. (2022). Parental and environmental control of seed dormancy in Arabidopsis thaliana. *Annual Review of Plant Biology, 73*, 355–378.

Jahnke, S., & Scholten, S. (2009). Epigenetic resetting of a gene imprinted in plant embryos. *Current Biology, 19*(19), 1677–1681.

Kermicle, J. L. (1970). Dependence of the R-mottled aleurone phenotype in maize on mode of sexual transmission. *Genetics, 66*, 69–85.

Khaitová, L. C., Fojtová, M., Křížová, K., Lunerová, J., Fulneček, J., Depicker, A., & Kovařík, A. (2011). Paramutation of tobacco transgenes by small RNA-mediated transcriptional gene silencing. *Epigenetics, 6*(5), 650–660.

Kinoshita, T., Miura, A., Choi, Y., Kinoshita, Y., Cao, X., Jacobsen, S. E., et al. (2004). One-way control of FWA imprinting in Arabidopsis endosperm by DNA methylation. *Science, 303*(5657), 521–523.

Kinoshita, Y., Saze, H., Kinoshita, T., Miura, A., Soppe, W. J., Koornneef, M., & Kakutani, T. (2007). Control of FWA gene silencing in Arabidopsis thaliana by SINE-related direct repeats. *The Plant Journal, 49*(1), 38–45.

Kirkbride, R. C., Lu, J., Zhang, C., Mosher, R. A., Baulcombe, D. C., & Chen, Z. J. (2019). Maternal small RNAs mediate spatial-temporal regulation of gene expression, imprinting, and seed development in Arabidopsis. *Proceedings of the National Academy of Sciences, 116*(7), 2761–2766.

Köhler, C., Hennig, L., Bouveret, R., Gheyselinck, J., Grossniklaus, U., & Gruissem, W. (2003). Arabidopsis MSI1 is a component of the MEA/FIE Polycomb group complex and required for seed development. *The EMBO Journal, 22*, 4804–4814.

Köhler, C., Page, D. R., Gagliardini, V., & Grossniklaus, U. (2005). The Arabidopsis thaliana MEDEA Polycomb group protein controls expression of PHERES1 by parental imprinting. *Nature Genetics, 37*(1), 28–30.

Law, J. A., & Jacobsen, S. E. (2010). Establishing, maintaining and modifying DNA methylation patterns in plants and animals. *Nature Reviews Genetics, 11*(3), 204–220.

Lind, M. I., & Spagopoulou, F. (2018). Evolutionary consequences of epigenetic inheritance. *Heredity, 121*(3), 205–209.

Lu, D., Zhai, J., & Xi, M. (2022). Regulation of DNA methylation during plant endosperm development. *Frontiers in Genetics, 13*, 760690.

Makarevich, G., Leroy, O., Akinci, U., Schubert, D., Clarenz, O., Goodrich, J., Grossniklaus, U., & Köhler, C. (2006). Different Polycomb group complexes regulate common target genes in Arabidopsis. *EMBO Reports, 7*, 947–952.

Makarevich, G., Villar, C. B., Erilova, A., & Köhler, C. (2008). Mechanism of PHERES1 imprinting in Arabidopsis. *Journal of Cell Science, 121*(6), 906–912.

McKeown, P. C., Fort, A., & Spillane, C. (2014). Analysis of genomic imprinting by quantitative allele-specific expression by pyrosequencing®. In C. Spillane & P. McKeown (Eds.), *Plant epigenetics and epigenomics. Methods in molecular biology* (Vol. 1112, pp. 85–104). Humana Press.

Meng, D., Zhao, J., Zhao, C., Luo, H., Xie, M., Liu, R., et al. (2018). Sequential gene activation and gene imprinting during early embryo development in maize. *The Plant Journal, 93*(3), 445–459.

Moreno-Romero, J., Del Toro-De León, G., Yadav, V. K., Santos-González, J., & Köhler, C. (2019). Epigenetic signatures associated with imprinted paternally expressed genes in the Arabidopsis endosperm. *Genome Biology, 20*, 1–11.

Pilu, R. (2015). Paramutation phenomena in plants. In *Seminars in cell & developmental biology* (Vol. 44, pp. 2–10). Academic.

Piskurewicz, U., Iwasaki, M., Susaki, D., Megies, C., Kinoshita, T., & Lopez-Molina, L. (2016). Dormancy-specific imprinting underlies maternal inheritance of seed dormancy in Arabidopsis thaliana. *eLife, 5*, e19573.

Roy, S., Gupta, P., Rajabhoj, M. P., Maruthachalam, R., & Nandi, A. K. (2018). The polycomb-group repressor MEDEA attenuates pathogen defense. *Plant Physiology, 177*(4), 1728–1742.

Satyaki, P. R. V., & Gehring, M. (2017). DNA methylation and imprinting in plants: Machinery and mechanisms. *Critical Reviews in Biochemistry and Molecular Biology, 52*(2), 163–175.

Selinger, D. A., & Chandler, V. L. (2001). B-Bolivia, an allele of the maize b1 gene with variable expression, contains a high copy retrotransposon-related sequence immediately upstream. *Plant Physiology, 125*, 1363–1379.

Soppe, W. J. J., Jacobsen, S. E., Alonso-Blanco, C., Jackson, J. P., Kakutani, T., Koornneef, M., & Peeters, A. J. M. (2000). The late flowering phenotype of fwa mutants is caused by gain-of-function epigenetic alleles of a homeodomain gene. *Molecular Cell, 2000*(6), 791–802.

Stam, M. (2009). Paramutation: A heritable change in gene expression by allelic interactions in trans. *Molecular Plant, 2*(4), 578–588.

Stam, M., Belele, C., Dorweiler, J. E., & Chandler, V. L. (2002). Differential chromatin structure within a tandem array 100 kb upstream of the maize b1 locus is associated with paramutation. *Genes & Development, 16*(15), 1906–1918.

TAIR (The Arabidopsis Information Resource). (2023). https://www.arabidopsis.org/gene?key= 429175, accded on July 2024.

Villar, C. B. R., Erilova, A., Makarevich, G., Trösch, R., & Köhler, C. (2009). Control of PHERES1 imprinting in Arabidopsis by direct tandem repeats. *Molecular Plant, 2*(4), 654–660.

Xin, M., Yang, R., Li, G., Chen, H., Laurie, J., Ma, C., et al. (2013). Dynamic expression of imprinted genes associates with maternally controlled nutrient allocation during maize endosperm development. *The Plant Cell, 25*(9), 3212–3227.

Yuan, J., Chen, S., Jiao, W., Wang, L., Wang, L., Ye, W., et al. (2017). Both maternally and paternally imprinted genes regulate seed development in rice. *New Phytologist, 216*(2), 373–387.

Zhang, M., Xie, S., Dong, X., Zhao, X., Zeng, B., Chen, J., et al. (2014). Genome-wide high resolution parental-specific DNA and histone methylation maps uncover patterns of imprinting regulation in maize. *Genome Research, 24*(1), 167–176.

Part III
Epigenetic Regulation as Mediator of Complex Genetic-Environment Interactions

Chapter 5
The RNA Interference Pathway

In plants, the conserved RNA interference (RNAi) pathway plays a pivotal role, functioning as both a robust defense mechanism against pathogens and a fine-tuned regulator of gene expression. RNAi has already shown its outstanding potential for agricultural applications. By engineering RNAi pathways, crop scientists can enhance pest and disease resistance, improve nutritional value, and optimize growth. These phenotypic changes can be attributed to siRNA/miRNA-mediated gene silencing during development.

RNA interference (RNAi) acts as a powerful defense mechanism against viruses and regulates gene expression. This intricate pathway is based on the ability of double-stranded RNA (dsRNA) molecules to trigger a silencing effect. RNAi involves small interfering RNAs (siRNAs) or microRNAs (miRNAs) that are incorporated into a protein complex referred to as the RNA-induced silencing complex (RISC). Additionally, long non-coding RNAs (lncRNAs) can also mediate RNA-transcriptional interference, specifically targeting the transcription of protein-coding genes. The RISC complex guides siRNAs/miRNAs to complementary sequences on target mRNA molecules, typically viral RNA or the mRNA of a specific plant gene. Once bound, the RISC complex can either cleave the target mRNA directly, leading to translational repression (reduced protein production), or mark it for degradation. This targeted silencing mechanism effectively prevents viral replication or down-regulates the expression of specific genes, allowing plants to fine-tune their response to internal and external cues.

The RNAi pathway comprises three core components, namely, small RNAs, Dicer-Like (DCL) endonucleases, and Argonaute (AGO) proteins. Small RNAs, encompassing miRNAs and siRNAs, are typically 20–24 nucleotides long and act as the guiding molecules in this pathway. DCL enzymes function as molecular scissors, processing double-stranded RNA precursors into small RNAs. AGO proteins, on the other hand, form the core of effector complexes that interact with small RNAs. These small RNA guides direct AGO proteins to complementary RNA targets, triggering either degradation or translational/transcriptional repression.

© The Author(s), under exclusive license to Springer Nature
Switzerland AG 2024
L. M. Vaschetto, *Epigenetics in Crop Improvement*,
https://doi.org/10.1007/978-3-031-73176-1_5

RNAi involves several sequential steps, including small RNA biogenesis, loading onto AGO components, and target recognition. Small RNA biogenesis initiates with the generation of double-stranded RNA (dsRNA) precursors, originating from various nucleotide sequence sources such as inverted repeats, endogenous gene transcripts, or RNA viruses (Morgado, 2020; Routhu et al., 2020).

An Overview on Small RNA Biogenesis

In plants, microRNAs (miRNAs) originate from MIR genes transcribed by RNA polymerase II into primary transcripts (pri-miRNAs) stem-loop structures containing both 5′ cap and 3′ poly A tails (Xie et al., 2005). miRNA biogenesis is initiated by DICER-LIKE1 (DCL1) processing pri-miRNAs into precursor miRNAs (pre-miRNAs). Plants typically encode four DCL enzymes with distinct functions. DCL1 specializes in microRNA (miRNA) biogenesis, while DCL2, DCL3, and DCL4 generate 22-, 24-, and 21-nucleotide small interfering RNAs (siRNAs), respectively (Qin et al., 2010; Wilson & Doudna, 2013). DCL1 cleaves pre-miRNAs to generate mature 21-nucleotide miRNA/miRNA duplexes, which is achieved in collaboration with HYPONASTY LEAVES1 (HYL1) and SERRATE (SE) proteins (Vazquez et al., 2004; Singh et al., 2018). Dicer-mediated cleavage precedes modification by HEN1 methyltransferase, which catalyzes the addition of a methyl group to the final nucleotide of each strand within miRNA duplexes (Yu et al., 2005). The miRNA duplex, containing the mature miRNA and its passenger strand, is then transported to the cytoplasm, where a helicase unwinds the duplex, allowing the mature miRNA to associate with the Argonaute (AGO1) protein within the RNA-induced silencing complex (RISC). The RISC-bound miRNA then guides target mRNA cleavage or translational repression (Guleria et al., 2011).

Unlike miRNAs, siRNAs originate from perfectly paired double-stranded RNAs (dsRNAs) arising from converging transcripts or RNA-dependent RNA polymerase activity (Bologna et al., 2018). DCL proteins then process these dsRNAs into siRNA duplexes via sequential cleavages. RNA Dependent RNA polymerases (RDRs) are also major protein components involved in siRNA biogenesis. Secondary siRNAs can be produced through an amplification process initiated by small RNA-guided cleavage of a target transcript, followed by RDR-mediated dsRNA synthesis (Carbonell, 2019). Figure 5.1 schematizes this feedback loop mechanism. For a detailed review on the molecular mechanisms involved in plant siRNA/miRNA biogenesis, please refer to Zhan and Meyers (2023), and Vaucheret and Voinnet (2024).

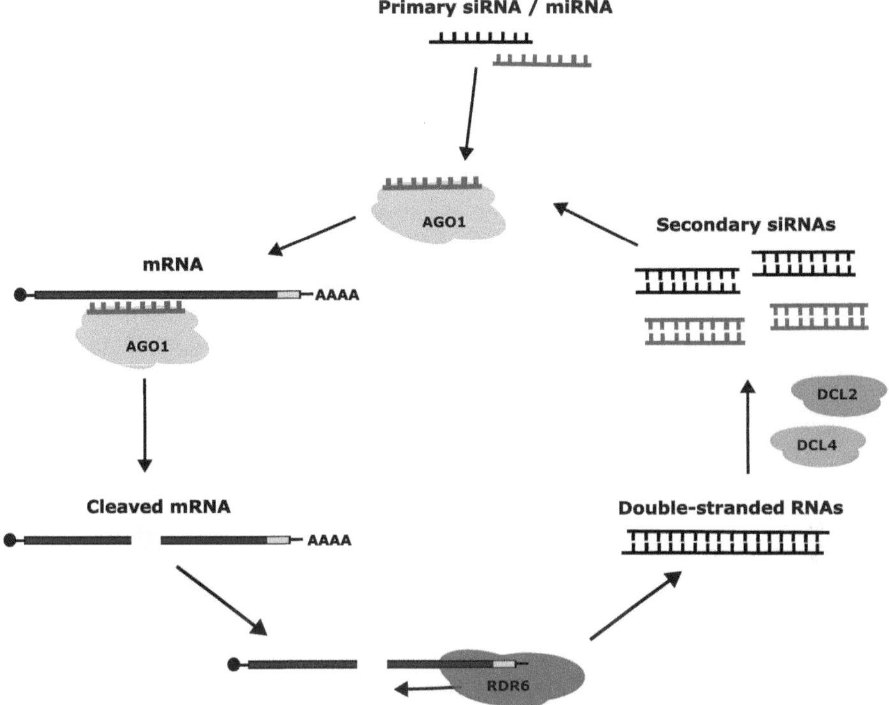

Fig. 5.1 The cycle of the RNA interference (RNAi) pathway. (Adapted from Hung & Slotkin, 2021). The evolutionary conserved Argonaute (AGO) protein family plays a pivotal role in RNAi through small-non coding RNA (sncRNA) binding. In *Arabidopsis*, AGO1 is a major member of this family, responsible for miRNA-guided gene silencing and also interacting with siRNAs. A RNAi amplification loop is established when AGO1, guided by primary siRNAs or miRNAs, cleaves target mRNAs. The RNA-dependent RNA Polymerase (RDR6) generates dsRNAs from these cleaved transcripts, which are processed into secondary siRNAs by Dicer-like 2 (DCL2) and DCL4. These secondary siRNAs then bind AGO1 to induce mRNA degradation, perpetuating the silencing cascade. Black circle—5′-cap; blue box—open reading frame; yellow box—terminator sequence

RNAi-Targeted Silencing

RNAi has emerged as a powerful tool for manipulating gene expression with remarkable precision. It harnesses the natural cellular machinery responsible for silencing genes, offering high precision for fine-tuning gene expression, allowing scientists to selectively knock down or silence specific target genes. This approach opens doors to exciting possibilities for enhancing crop traits. One of the most compelling applications of RNAi in agriculture lies in combating multiple crop pathogens, including viruses, fungi, bacteria, and nematodes (Vaschetto, 2021; Hernández-Soto & Chacón-Cerdas, 2021). By identifying and silencing exogenous genes, researchers can effectively engineer crops with inherent resistance to detrimental pathogenic invaders. This targeted approach offers several advantages over traditional methods of disease control:

- Reduced reliance on chemical pesticides: RNAi-based resistance can significantly decrease the need for chemical applications, promoting sustainable and environmentally friendly agricultural practices (Vaschetto & Beccacece, 2019).
- Broad-spectrum protection: Targeting essential pathogen genes provides broad-spectrum resistance against a range of pathogens within the same family, offering long-term protection and reducing the risk of resistance development (Ghag, 2017).
- Enhanced durability: Silencing essential genes within the pathogen leads to more durable resistance, minimizing the likelihood of the pathogen evolving to overcome the defense mechanism (Saakre et al., 2023).

RNAi: A Tool for Environmental Stress Resilience

RNAi has emerged as a powerful tool for enhancing stress tolerance. By silencing negative regulators and amplifying stress-responsive genes, RNAi reprograms gene expression, enabling plants to better adapt to environmental changes, including drought, salinity, and extreme temperatures. Hence, RNAi-mediated silencing of target genes involved in water loss -or activating genes associated with water absorption- can significantly improve plant ability to withstand drought conditions. In tomato plants, Mushtaq et al. (2022) demonstrated improved drought tolerance through RNAi-mediated downregulation of SlHK2, a cytokinin receptor gene. This manipulation enhanced membrane stability, osmoprotectant accumulation, and antioxidant enzyme activities. In rice, Manavalan et al. (2012) employed RNAi to disrupt the *farnesyltransferase/squalene synthase (SQS)* gene using maize squalene synthase, leading to improved drought tolerance in transgenic rice plants at both vegetative and reproductive stages. Remarkably, transgenic plants displayed enhanced recovery and increased grain yield upon rehydration. They observed that RNAi-mediated *SQS* inactivation resulted in reduced stomatal conductance, a key factor contributing to drought tolerance (Manavalan et al., 2012). Also in rice, Li et al. (2009) investigated the role of the *RACK1* gene in drought tolerance using RNA interference (RNAi). They found that downregulation of *RACK1* expression through RNAi enhanced drought tolerance in transgenic rice plants compared to non-transgenic controls. This suggests that RACK1 functions as a negative regulator of the redox system-mediated drought stress response in rice (Li et al., 2009).

RNAi targeting genes that regulate salt accumulation and activate stress-protective pathways can also help to develop crops with increased salinity tolerance. By exploiting RNAi technology, Zhang et al. (2018) characterized the roles of Ca2 + -dependent protein kinases (CDPKs) in *Arabidopsis* salt stress tolerance. Their *CPK12-RNAi* mutant plants displayed increased salt sensitivity during seedling growth compared to wild-type plants. This suggests CPK12 mediates salt tolerance by regulating ion homeostasis and hydrogen peroxide (H_2O_2) production. Liu et al. (2022) developed a promising approach for enhancing rice yield under saline-alkaline (SA) stress conditions. By employing RNAi to suppress *OsABA8ox1*, a key ABA catabolic gene, they generated *OsABA8ox1-kd* plants exhibiting increased endogenous ABA content and improved tolerance to SA stress. Notably, these lines displayed a significant increase in grain yield compared to wild-type plants. These findings suggest that manipulating ABA levels through *OsABA8ox1* suppression offers a novel molecular strategy for improving rice productivity in high SA conditions (Liu et al., 2022). In rice seedlings, Lin et al. (2024) revealed that transgenic *OsBBTI5*-RNAi plants exhibit enhanced stress tolerance by modulating H_2O_2 levels and gibberellic acid (GA3) signaling. *OsBBTI5* encodes a protease inhibitor referred to as Bowman-Birk inhibitor (BBI) that plays roles in plant stress responses.

High temperature is a major environmental stressor negatively impacting crop yields and a major objective when developing crop varieties resilient to climate change. Zhao et al. (2020) explored RNAi as a strategy to address heat stress responses. They developed dsRNA interference vectors targeting the *OsSSIIIa* gene and introduced them into the rice genome, generating transgenic lines. These *SSIIIa-RNAi* plants displayed increased susceptibility to grain chalkiness under high temperatures compared to controls. This effect was attributed to the reduced activity of AGPase, an enzyme essential for starch biosynthesis (Zhao et al., 2020). Marín-Sanz et al. (2020) investigated the effects of temperature stress and nitrogen availability on wheat lines engineered with RNAi-mediated gliadin downregulation. While grain weight decreased under heat stress in all lines, gliadin content only increased in wild-type plants. These findings suggest that RNAi lines maintain stable gliadin silencing despite environmental challenges, and they therefore appear well-suited for producing wheat grain with low gliadin content, particularly under conditions of rising temperatures and variable nitrogen availability (Marín-Sanz et al., 2020).

RNAi also offers a promising alternative for improving the nutritional value of crops by reducing anti-nutritional compounds and enhancing essential nutrient content. This tool has already been shown to be effective for silencing genes responsible for the production of compounds like anti-nutritional phytates, which inhibit nutrient absorption. Kumar et al. (2019) utilized RNAi to target *GmMIPS1* in soybean seeds, demonstrating reduced phytate accumulation and enhanced mineral bioavailability. Similarly, RNAi targeting genes involved in nutrient transport and storage can increase the levels of essential nutrients in crops (Tang & Galili, 2004).

The emerging link between non-coding RNAs (ncRNAs) and responses to temperature stress, particularly heat stress, opens avenues for applying the RNAi technology under challenging climate conditions. Temperature stress induces fluctuations in ncRNA abundance, leading to altered levels of their target mRNAs, protein synthesis, and subsequent stress responses. Jin et al. (2024) identified over 500 miRNAs and 1300 miRNA-mRNA interactions implicated in jujube leaf response to high-temperature stress. Moreover, Ci et al. (2015) reported over 1000 differentially methylated sites in *Populus simonii*, a species of poplar native to northeast China and to Mongolia, under temperature stress. They suggested that DNA methylation influences miRNA gene expression, thereby modulating target gene activity via miRNA-mediated gene silencing to ensure cellular viability during abiotic stress. Hence, by manipulating ncRNA expression through RNAi, we can potentially modulate plant stress responses and engineer climate-resilient crops. Zhong et al. (2013) unveiled a link between warming temperatures and siRNA-mediated epigenetic regulation in *Arabidopsis*. Their study revealed that warm temperatures trigger a transgenerational release of RNA silencing through inhibition of siRNA biogenesis. This effect stemmed from a temperature-induced decrease in SUPPRESSOR OF GENE SILENCING 3 (SGS3) protein abundance, a key factor for stable double-stranded RNA (dsRNA) formation. *SGS3* overexpression rescued the warmth-triggered suppression of siRNA biogenesis and diminished the transgenerational epigenetic memory (Zhong et al., 2013). Moreover, Song et al. (2020) identified high-temperature-responsive poplar lncRNAs functioning as regulators of heat tolerance. These lncRNAs modulate target gene expression through either RNA interference or RNA scaffolding mechanisms. Remarkably, overexpressing specific lncRNA targets in *Arabidopsis* enhanced heat tolerance, highlighting the potential for manipulating lncRNAs to improve plant stress resilience (Song et al., 2020).

Potential Limitations of the RNAi Technology

One of the primary concerns associated with RNAi-based strategies is the unintentional silencing of genes in target and non-target organisms. This can occur when the RNAi molecule or synthetically designed siRNAs share sequence similarity with other genes within the genome. If the siRNA binds to and degrades mRNA molecules of non-target genes, it can result in unintended changes in gene expression (Senthil-Kumar & Mysore, 2011). In some instances, RNAi molecules produced by genetically modified crops might be released into the environment through pollen. If these RNAi molecules are taken up by non-target organisms, such as beneficial insects or soil microorganisms, they may then potentially impact the gene expression and physiology of these species. It is only one example why it is important to take careful consideration of potential off-target effects when designing and using RNAi technologies in agriculture. Beyond target specificity, developing efficient delivery systems that ensure tissue-specific gene silencing also remains a major challenge for optimizing RNAi technology.

Conclusions

As climate change disrupts weather patterns and elevates temperatures, ensuring crop resilience to abiotic stresses like drought, salinity, and heat becomes a critical issue in molecular breeding programs. The RNAi technology is an powerful tool to silence target genes that hinder stress tolerance and amplify the expression of stress-responsive genes. This allows plants to better cope with challenging environmental conditions. While challenges remain in efficiently delivering RNA molecules and ensuring target specificity, RNAi technology holds immense promise for developing climate-resilient crops. By strategically manipulating gene expression through RNAi, we can engineer crops better equipped to fight against environmental stresses associated with climate change, ultimately contributing to food security.

References

Bologna, N. G., Iselin, R., Abriata, L. A., Sarazin, A., Pumplin, N., Jay, F., et al. (2018). Nucleo-cytosolic shuttling of ARGONAUTE1 prompts a revised model of the plant microRNA pathway. *Molecular Cell, 69*(4), 709–719.

Carbonell, A. (2019). Secondary small interfering RNA-based silencing tools in plants: An update. *Frontiers in Plant Science, 10*, 687.

Ci, D., Song, Y., Tian, M., & Zhang, D. (2015). Methylation of miRNA genes in the response to temperature stress in Populus simonii. *Frontiers in Plant Science, 6*, 921.

Ghag, S. B. (2017). Host induced gene silencing, an emerging science to engineer crop resistance against harmful plant pathogens. *Physiological and Molecular Plant Pathology, 100*, 242–254.

Guleria, P., Mahajan, M., Bhardwaj, J., & Yadav, S. K. (2011). Plant small RNAs: Biogenesis, mode of action and their roles in abiotic stresses. *Genomics, Proteomics & Bioinformatics, 9*(6), 183–199.

Hernández-Soto, A., & Chacón-Cerdas, R. (2021). RNAi crop protection advances. *International Journal of Molecular Sciences, 22*(22), 12148.

Hung, Y. H., & Slotkin, R. K. (2021). The initiation of RNA interference (RNAi) in plants. *Current Opinion in Plant Biology, 61*, 102014.

Jin, J., Yang, L., Fan, D., Li, L., & Hao, Q. (2024). Integration analysis of miRNA-mRNA pairs between two contrasting genotypes reveals the molecular mechanism of jujube (Ziziphus jujuba Mill.) response to high-temperature stress. *BMC Plant Biology, 24*(1), 612.

Kumar, A., Kumar, V., Krishnan, V., Hada, A., Marathe, A., Jolly, M., & Sachdev, A. (2019). Seed targeted RNAi-mediated silencing of GmMIPS1 limits phytate accumulation and improves mineral bioavailability in soybean. *Scientific Reports, 9*(1), 7744.

Li, D. H., Hui, L. I. U., Yang, Y. L., Zhen, P. P., & Liang, J. S. (2009). Down-regulated expression of RACK1 gene by RNA interference enhances drought tolerance in rice. *Rice Science, 16*(1), 14–20.

Lin, Z., Yi, X., Ali, M. M., Zhang, L., Wang, S., Tian, S., & Chen, F. (2024). RNAi-mediated suppression of OsBBTI5 promotes salt stress tolerance in Rice. *International Journal of Molecular Sciences, 25*(2), 1284.

Liu, X., Xie, X., Zheng, C., Wei, L., Li, X., Jin, Y., et al. (2022). RNAi-mediated suppression of the abscisic acid catabolism gene OsABA8ox1 increases abscisic acid content and tolerance to saline–alkaline stress in rice (Oryza sativa L.). *The Crop Journal, 10*(2), 354–367.

Manavalan, L. P., Chen, X., Clarke, J., Salmeron, J., & Nguyen, H. T. (2012). RNAi-mediated disruption of squalene synthase improves drought tolerance and yield in rice. *Journal of Experimental Botany, 63*(1), 163–175.

Marín-Sanz, M., Giménez, M. J., Barro, F., & Savin, R. (2020). Prolamin content and grain weight in RNAi silenced wheat lines under different conditions of temperature and nitrogen availability. *Frontiers in Plant Science, 11*, 511647.

Morgado, L. (2020). Biogenesis of small RNA: Molecular pathways and regulatory mechanisms. In *Plant small RNA* (pp. 49–64). Academic.

Mushtaq, N., Wang, Y., Fan, J., Li, Y., & Ding, J. (2022). Down-regulation of cytokinin receptor gene SlHK2 improves plant tolerance to drought, heat, and combined stresses in tomato. *Plants, 11*(2), 154.

Qin, H., Chen, F., Huan, X., Machida, S., Song, J., & Yuan, Y. A. (2010). Structure of the Arabidopsis thaliana DCL4 DUF283 domain reveals a noncanonical double-stranded RNA-binding fold for protein-protein interaction. *RNA, 16*(3), 474–481.

Routhu, G., Borah, M., Nath, P. D., & Deb, B. (2020). RNA interference (RNAi) and response of plant cells to double stranded RNA (dsRNA). *International Journal of Current Microbiology and Applied Sciences, 9*(9), 3114–3125.

Saakre, M., Jaiswal, S., Rathinam, M., Raman, K. V., Tilgam, J., Paul, K., et al. (2023). Host-delivered RNA interference for durable pest resistance in plants: Advanced methods, challenges, and applications. *Molecular Biotechnology, 66*, 1–20.

Senthil-Kumar, M., & Mysore, K. S. (2011). Caveat of RNAi in plants: The off-target effect. In H. Kodama & A. Komamine (Eds.), *RNAi and plant gene function analysis. Methods in molecular biology* (Vol. 744, pp. 13–25). Humana Press.

Singh, A., Gautam, V., Singh, S., Sarkar Das, S., Verma, S., Mishra, V., et al. (2018). Plant small RNAs: Advancement in the understanding of biogenesis and role in plant development. *Planta, 248*, 545–558.

Song, Y., Chen, P., Liu, P., Bu, C., & Zhang, D. (2020). High-temperature-responsive poplar lncRNAs modulate target gene expression via RNA interference and act as RNA scaffolds to enhance heat tolerance. *International Journal of Molecular Sciences, 21*(18), 6808.

Tang, G., & Galili, G. (2004). Using RNAi to improve plant nutritional value: From mechanism to application. *Trends in Biotechnology, 22*(9), 463–469.

Vaschetto, L. M. (2021). *RNAi strategies for pest management. Methods and protocols.* Springer Nature.

Vaschetto, L. M., & Beccacece, H. M. (2019). The emerging importance of noncoding RNAs in the insecticide tolerance, with special emphasis on Plutella xylostella (Lepidoptera: Plutellidae). *Wiley Interdisciplinary Reviews: RNA, 10*(5), e1539.

Vaucheret, H., & Voinnet, O. (2024). The plant siRNA landscape. *The Plant Cell, 36*(2), 246–275.

Vazquez, F., Gasciolli, V., Crete, P., & Vaucheret, H. (2004). The nuclear dsRNA binding protein HYL1 is required for microRNA accu- mulation and plant development, but not post-transcriptional transgene silencing. *Current Biology, 14*(4), 346–351. https://doi.org/10.1016/j.cub.2004.01.035

Wilson, R. C., & Doudna, J. A. (2013). Molecular mechanisms of RNA interference. *Annual Review of Biophysics, 42*, 217–239.

Xie, Z., Allen, E., Fahlgren, N., Calamar, A., Givan, S. A., & Carrington, J. C. (2005). Expression of Arabidopsis MIRNA genes. *Plant Physiology, 138*(4), 2145–2154. https://doi.org/10.1104/pp.105.062943

Yu, B., Yang, Z., Li, J., Minakhina, S., Yang, M., Padgett, R. W., et al. (2005). Methylation as a crucial step in plant microRNA biogenesis. *Science, 307*(5711), 932–935.

Zhan, J., & Meyers, B. C. (2023). Plant small RNAs: Their biogenesis, regulatory roles, and functions. *Annual Review of Plant Biology, 74*(1), 21–51.

Zhang, H., Zhang, Y., Deng, C., Deng, S., Li, N., Zhao, C., et al. (2018). The Arabidopsis Ca2+-dependent protein kinase CPK12 is involved in plant response to salt stress. *International Journal of Molecular Sciences, 19*(12), 4062.

Zhao, Q., Ye, Y., Han, Z., Zhou, L., Guan, X., Pan, G., & Cheng, F. (2020). SSIIIa-RNAi suppression associated changes in rice grain quality and starch biosynthesis metabolism in response to high temperature. *Plant Science, 294*, 110443.

Zhong, S. H., Liu, J. Z., Jin, H., Lin, L., Li, Q., Chen, Y., et al. (2013). Warm temperatures induce transgenerational epigenetic release of RNA silencing by inhibiting siRNA biogenesis in Arabidopsis. *Proceedings of the National Academy of Sciences, 110*(22), 9171–9176.

Chapter 6
The Role of Transposable Elements in Plant Development

The plant genome is significantly influenced by the presence and activity of mobile genetic sequences known as transposable elements (TEs). These sequences possess the ability to relocate and replicate within the host genome, contributing to both genetic variation and evolutionary processes. Plant TEs exhibit a significant diversity, from their modes of transposition and activity levels to their species-specific insertion preferences. This dynamism fuels genetic and epigenetic variation and shapes plant genome evolution through redundancy and chromosome rearrangements. TE activity exerts multiple effects on plant genomes, including gene disruption, *de novo* gene function through TE exaptation, and modulation of gene expression. By assessing TE activity, researchers can gain insights into how these mobile elements influence epigenetic landscapes, ultimately shaping gene expression patterns in economically important crops.

Classification of Transposable Elements and Functional Importance

Transposable elements (TEs) fall into two major classes based on their transposition mechanisms: retrotransposons (Class I) and DNA transposons (Class II). These classes exhibit distinctive characteristics and modes of action, ultimately leading to diverse functional consequences. Retrotransposons mobilize via an RNA intermediate ("copy and paste"), integrating at new *loci* and impacting genome size and gene expression (Chénais et al., 2012; Schrader & Schmitz, 2019). DNA transposons, on the other hand, employ a "cut and paste" mechanism, directly excising and reintegrating within the genome, leading to significant alterations in architectural organization and functionality (Feschotte & Pritham, 2007).

TE mobilization significantly impacts gene expression by inducing epigenetic changes on both *trans-* and *cis*-regulatory sequences, leading to altered

L. M. Vaschetto, *Epigenetics in Crop Improvement*,
https://doi.org/10.1007/978-3-031-73176-1_6

transcriptional activity (Fultz et al., 2015). Additionally, mobile elements contribute to shaping global gene expression patterns through the establishment of repressive epigenetic marks. The vast population of silenced TEs within crop genomes presents a valuable tool for targeting their active counterparts via RNA-directed DNA methylation (RdDM) pathways.

Epigenetic silencing represents the primary defense mechanism against TE activity, safeguarding genomic stability and integrity. This silencing involves the accumulation of repressive epigenetic marks on TEs, leading to transcriptional inhibition and potentially impacting neighboring genes or even entire linkage groups (Vaschetto, 2016). TE mobilization triggers the recruitment of the epigenetic machinery, ultimately triggering their repression through the deposition of suppressive marks like histone modifications and DNA methylation. Additionally, TEs often harbor regulatory elements within their nucleotide sequences. Consequently, their integration can inadvertently modulate the expression of adjacent genes (Hirsch & Springer, 2017). Deciphering the regulatory networks controlling TE activity holds immense potential for targeted crop improvement. While TE detection and analysis remain challenging due to their repetitive sequence nature, innovative bioinformatic approaches are currently providing valuable insights into TE movement and behavior across diverse genomic scales.

Transposable Elements and Genome Size

Plant TEs exhibit remarkable variation in abundance and diversity between species. These fluctuations significantly contribute to the observed disparities in plant genome size, underscoring the dynamic interplay between TEs and their host genomes. Crop genomes, for instance, can harbor a vast majority of TEs, constituting up to 90% of their genetic makeup (Vicient, 2010). These dynamic sequences can be categorized as either relics of past transposition events or active elements, identifiable by recent bursts of expansion within the genome. The amplification of TEs is driven by their unique transposition mechanisms. Class I TEs directly contribute to their copy number expansion through a characteristic copy-and-paste mode of transposition. By transcribing their genomic sequence into RNA and then reverse-transcribing it back into DNA for integration at new genomic locations, they essentially replicate themselves within the host genome. Conversely, Class II TEs excise themselves from one genomic location and reinsert elsewhere, dependent on host-mediated repair at the original site, which does not directly increase their copy number.

Given their sequence similarity, TEs can be organized into families, revealing a shared lineage for members within each group. Some TE families experience rapid TE amplification (bursts) events across taxonomic groups, drastically reshaping the genetic landscape through their proliferating copies. Movement and amplification of TEs is triggered by transposase enzymes for transposons and the reverse transcription machinery for retrotransposons. Environmental stresses (either biotic or abiotic stress) can activate TEs, potentially increasing transposition rates and providing an immediate source of variation for adaptation (Makarevitch et al., 2015). TE amplification can lead to exaptation, an evolutionary process where TEs acquire novel functions like gene regulation or even gene formation (Joly-Lopez et al., 2016). These exapted genes can persist by providing a selective advantage to the organism, highlighting the potential for TEs to contribute to functional innovation.

TEs and their derived sequences act as epigenetic regulators of gene expression, playing a key role in gene function. One such mechanism involves proximity effects, where the physical proximity of mobile sequences influences neighboring genes by altering chromatin structure. These sequences can also establish scaffold/matrix attachment regions (S/MARs) that mediate chromosomal loops and define chromatin domains, facilitating coordinated control of distant genomic regions (Pathak et al., 2024). The interplay between TEs and S/MARs suggests a potential mechanism for shaping genome structure and driving evolutionary change. On the other hand, TEs and TE-derived repeat sequences can function as hotspots for epigenetic modifications, potentially influencing gene expression (Vaschetto, 2016). Remarkably, TE mobilization events frequently occur in response to stress conditions. Upon TE insertion, these bursts of amplification can lead to the silencing or activation of multiple genes. A common TE-induced silencing mechanism involves DNA methylation spreading to nearby genes, resulting in reduced expression (Galindo-González et al., 2018). Alternatively, TEs can directly impact the expression of nearby genes in other ways. For instance, TEs may harbor regulatory DNA sequences, such as enhancers or promoters. Moreover, TEs may encode non-coding RNAs (ncRNAs) that modulate gene expression at both the post-transcriptional and transcriptional levels (Vaschetto, 2016). Transposition events can also potentially lead to gene inactivation or loss of function if TE insertion disrupts a gene coding sequence (Liu et al., 2014). As mentioned before, TEs can become exapted as new exons, leading to alternative splicing and the creation of new gene isoforms (Kim et al., 2022). This last process holds potential to increase protein diversity and functional complexity within host organisms.

Epigenetic Regulation of Transposable Elements

The pivotal role of epigenetic mechanisms in governing TE activity within the genome has been well established over the last decades. Disruption of epigenetic pathways have profound consequences on TE activity, ultimately leading to genomic instability and altered gene regulation. DNA methylation plays a crucial role in maintaining TE quiescence and ensuring genomic integrity. Disruption of this epigenetic mark can lead to TE activation and potentially deleterious effects on gene expression. Extensive DNA methylation patterns within TEs serve as a potent silencing mechanism, ensuring their stable inheritance through mitotic cell divisions. However, TE demethylation may lead to their reactivation and potential deleterious transposition events, thereby compromising genomic integrity. Histone modifications constitute an additional layer of epigenetic control over TEs. Specific marks associated with active or repressed chromatin states can exert a profound influence on TE activity, either silencing them or facilitating their mobilization. Small RNAs also contribute to TE silencing through the RdDM pathway. The interaction between ncRNAs with AGO proteins directs the recruitment of chromatin modifiers, leading to the deposition of repressive epigenetic marks (e,g, H3K9me and DNA methylation) at TE *loci*, ensuring continued silencing and maintaining genomic stability (Fultz et al., 2015).

Transposable Elements in Crop Improvement

TEs can exert beneficial effects on agronomically important traits like stress tolerance, yield, or plant architectural changes, making them prime candidates for integration into the host genome as functional elements. This concept, known as TE domestication, involves identifying and exploiting beneficial TEs to unlock their potential for increasing crop performance (Thieme & Bucher, 2018). The insertion of the *Hopscotch* transposon within the regulatory domain of the *teosinte branched 1* (*tb1*) gene stands as a primary example of TEs as drivers of phenotypic evolution. This event significantly increased *tb1* expression, leading to the characteristic morphological structure of the maize plant (Studer et al., 2011).

The inherent activity of TEs has been repurposed as a powerful instrument for generating genetic diversity in crop plants through the process of TE-induced mutagenesis (Zhou et al., 2016; Ulukapi & Nasircilar, 2018). This strategy entails the controlled activation of transposition within the plant genome, effectively harnessing the mutagenic nature of these mobile elements. Targeted TE insertion can generate potentially heritable mutations and enrich the genetic pool of crop populations, paving the way for the development of improved traits and enhanced agricultural resilience. By harnessing TE mobility, molecular breeders can induce targeted insertional mutations in genes and/or their regulatory regions, ultimately leading to new traits and altered gene functions. As a result, this approach offers a valuable tool for molecular crop improvement. Moreover, TE-induced mutagenesis also potentiates the discovery of *loci* associated with crop traits. By triggering mutations on a high scale, researchers can efficiently screen for desirable phenotypic changes. This approach has dramatically accelerated the process of trait discovery, offering an invaluable resource for molecular breeders.

Gene Regulation and TE Activity

TEs play multiple roles in shaping plant chromosomes, acting as both local and global regulators of gene expression. TE-derived enhancers can function as *cis*-regulatory elements, directly influencing gene expression at the site of transcription.

Moreover, overexpression of genes harboring TE-like enhancers can lead to their function as *trans*-regulators, controlling the expression of genes downstream in critical developmental pathways. Similarly, TE-mediated epigenetic regulation serves as a central control mechanism within signaling pathways. TEs and their derived sequences function as epigenetic control components, interacting with genes (genetic integrators) located within the same regulatory *loci*. This interplay exemplifies the independent yet complementary roles of genetic and epigenetic mechanisms in shaping plant development. A prime case of such a synergistic effect acting on master developmental pathways is *tb1* in maize. This *locus* plays a crucial role in repressing lateral branching, contributing to the more compact architecture observed in modern maize plants compared to their wild antecessor teosinte. The *tb1 locus* underwent the insertion of a TE (*Hopscotch*) within its regulatory enhancer region, upregulating its expression and accounting for the enhanced apical dominance observed in maize compared to its closest relative teosinte (Studer et al., 2011).

Epigenetic phenomena like paramutation, a meiotically and mitotically heritable epigenetic process, can be instrumental in elucidating the role of TEs and repetitive (often TE-derived) sequences in mediating epigenetically inherited changes. The *booster 1* (*b1*) *locus* in maize is a well-characterized case of paramutation that exemplifies this phenomenon. Genetically identical *b1* alleles exhibit distinct epigenetic states, termed epialleles. This *locus* harbors seven upstream tandem repeats acting as enhancers, positioned at a similar distance to those observed for the regulatory enhancer region of the *tb1* start codon (approximately 60 kb and 100 kb upstream start codons for *tb1* and *b1* genes, respectively). The tandem repeats at *b1* are associated with transgenerational epigenetic modulation, thereby evidencing a role for the nucleotide sequence context of the upstream region in establishing epigenetic states that exert regulatory effects through epigenetic modifications. TEs may also potentially serve as control elements, which has shown to be particularly important in major developmental *loci* like *tb1*. These elements can enhance expression of downstream genes through canonical genetic mechanisms (e.g., transcription factor binding), harbor nucleotide sequences that regulate gene expression at transcriptional level (e.g., ncRNAs) and serve as target repeat sequences for epigenetic modifications (e.g., DNA methylation), ultimately leading to changes in gene expression.

Chromatin remodeling complexes (CRCs) play key roles in gene regulation by bridging transcription factors and gene expression. Their activity is often coordinated with non-coding RNAs (ncRNAs) produced either locally (*cis*) or from distant (*trans*) *loci* (Patty & Hainer, 2020). While inverted repeat sequences are recognized as the most efficient triggers for dsRNA-mediated silencing, directly repeated and even single-copy sequences can also generate dsRNAs. However, inverted repeats promote more stable silencing due to efficient dsRNA production (Louwers et al., 2005). Conversely, silencing initiated by direct repeats relies on less frequent antisense RNA formation and exhibits lower persistence (Martienssen, 2003). This finding highlights the link between nucleotide sequence contexts and the resulting epigenetic modifications associated with changes in gene expression.

Inverted repeat-induced epigenetic changes might be reversible, while those triggered by direct repeats may lead to more persistent regulation.

As indicated above, TEs can harbor ncRNAs capable of inducing epigenetic modifications, ultimately regulating gene expression. It has already been suggested that *tb1* overexpression may involve TE-derived ncRNAs (Clark et al. (2006), potentially modifying chromatin accessibility. Reinforcing this hypothesis, Wang et al. (2022) identified an antisense lncRNA MSTRG.28151.1 encoded by the *tb1* orthologue of tobacco *NtTB1*. Interestingly, silencing this lncRNA significantly reduced *NtTB1* expression and resulted in larger axillary buds, highlighting a similar role to maize in plant development.

TEs and their derived sequences may serve as hotspots for linking multiple epigenetic mechanisms. Miniature inverted-repeat transposable elements (MITEs) are ubiquitous mobile elements that exemplify this concept. As the most active TEs in plant genomes, MITEs contribute to diverse phenotypic outcomes (Casacuberta & Santiago, 2003; Barteri, 2019). Tikhonov et al. (2000) demonstrated co-localization of MITEs and matrix attachment regions (MARs) on genomic fragments in maize and sorghum, suggesting their potential role in chromatin architecture. *In vitro* binding to isolated nuclear matrices suggests a potential role for MITEs as functional MAR elements *in vivo* (Tikhonov et al., 2000). The tandem and inverted repeat sequences within MITE sequences likely facilitate loop formation and serve as binding sites for chromatin remodeling complexes (Vaschetto, 2016). Additionally, MITE-encoded ncRNAs have shown to play crucial roles in gene regulation and even have contributed to miRNA gene evolution (Pegler et al., 2023). This emerging perspective not only challenges the traditional view of TEs as solely parasitic elements but instead positions them as fundamental partners in gene expression through multiple epigenetic mechanisms.

Epigenetic Targets of Major Developmental Genes

Teosinte branched 1 (*tb1*) and *Reduced Height* (*Rht*) are master regulatory genes of maize and bread wheat, respectively, and they govern multiple signaling pathways critical for agronomic yield. Unraveling epigenetic mechanisms controlling these essential developmental genes holds tremendous potential for molecular breeding of complex traits. Their epigenetic targeting offers an invaluable strategy for improving multiple crop traits without altering DNA sequence, minimizing disruptions to secondary signaling pathways.

Tb1

Although it has been shown that the *tb1* gene played a critical role in maize domestication, precise molecular mechanisms controlling its expression remains elusive (Clark et al., 2004; Vaschetto, 2015). It encodes a TCP transcription factor, Teosinte Branched 1 (Tb1), which regulates meristem growth, floral initiation, cell cycle, and differentiation (Doebley et al., 1997). Remarkably, *tb1* expression is doubled in maize compared to its closest relative teosinte, ultimately suppressing branching while promoting female inflorescence formation (Clark et al., 2006). Intriguingly, a ~ 12 kb enhancer region upstream of *tb1* (58–70 kb) is the key to its overexpression. This region contains TE insertions specific to maize and absent in teosinte (Zhou et al., 2011; Studer et al., 2011; Studer & Doebley, 2012; Prakash et al., 2020). As described above, Clark et al. (2006) hypothesized the presence of an unidentified element, potentially a functional RNA species, within this upstream region. This putative element might influence *tb1* transcript levels, ultimately contributing to phenotypic variation.

Cahn et al. (2024) recently identified a comprehensive set of regulatory regions in maize inbred lines, including distal enhancers distinguished from gene bodies by the absence of H3K4me1. Interestingly, one of these enhancer elements is associated with *tb1* and expressed non-coding enhancer RNAs (ncRNAs) bi-directionally. Bidirectional transcription-derived dsRNA serves as a primary template for generating regulatory small RNAs that modulate gene expression at both transcriptional and post-transcriptional levels, with siRNAs originating from tandem repeats in the paramutagenic *B′* allele of the *b1* gene (Sidorenko et al., 2009; Arteaga-Vazquez et al., 2010) exemplifying this process. These findings parallel the effects observed with the hepta-repeat enhancer function at the *b1 locus*. Hövel et al. (2024) demonstrated that mutations in key RdDM components, including *mop1* and *mop3*, led to reduced H3K9me2 and H3K27me2 levels at the hepta-repeat *locus*. Concomitantly, they observed partial activation of the hepta-repeat enhancer, altered chromatin architecture, and increased *b1* gene expression. Delineating the epigenetic mechanisms underlying the activity of TE and repeat, often TE-derived, sequences will significantly advance our understanding of how epigenetic signaling cascades act in parallel to genetic pathways to shape complex crop traits.

Rht1 *Genes*

The introgression of *Rht-B1b* and *Rht-D1b* major developmental genes into hexaploid bread wheat (*Triticum aestivum L.*) significantly boosted yields during the Green Revolution of the 1960s. *Rht* dwarfing genes, physically mapped to the short arms of chromosomes 4B (*Rht-B1*) and 4D (*Rht-D1*), encode truncated DELLA proteins that are insensitive to gibberellin (GA) signaling, which acts as a primary input, stimulating DELLA degradation via the 26S proteasome. These gain-of-function mutations effectively inhibit vegetative growth and GA responses, leading to reduced plant height and a reallocation of resources towards grain development (Hedden, 2003). Remarkably, it has already been shown that DELLA proteins function at the intersection of genetic and epigenetic mechanisms of gene expression control during development, ultimately modulating global transcriptional responses. In *Arabidopsis*, Sarnowska et al. (2013) revealed an interaction network between DELLA proteins and SWI/SNF chromatin remodeling complexes, which regulate transcription, DNA processes, and the cell cycle. Moreover, Huang et al. (2023) observed direct interactions between DELLA proteins and hundreds of transcription factors. These authors demonstrated that the DELLA protein REPRESSOR OF ga1–3 (RGA) binds histone H2A, forming complexes with hundreds of different transcription factors at target chromatin sites (Huang et al., 2023). DELLA stability is regulated by multiple factors beyond GA signaling, including alternative degradation pathways and post-translational modifications such as phosphorylation, SUMOylation, and glycosylation (Blanco-Touriñán et al., 2020).

The polyploid nature of hexaploid bread wheat renders it an exceptional model for investigating the intricate interplay between genetic and epigenetic mechanisms in regulating key *loci*. Polyploidy involves genome redundancy and therefore necessitates a more precise control of gene expression to prevent functional overlap. Wilhelm et al. (2013) observed a high evolutionary conservation among *Rht-1* sequences and their upstream flanking genes across A, B, and D wheat genomes, while TE insertions, except for MITEs, showed low conservation. The high sequence conservation observed within *Rht-1* sequences and their flanking regions in hexaploid wheat suggests a pivotal role for epigenetic mechanisms in the concerted regulation of *Rht-1* genes. Delving into the interplay between epigenetic modifications within conserved sequences, including MITE insertions, holds promise for understanding how *Rht-1* genes orchestrally act to regulate developmental processes. Elucidating the intersection between epigenetic and genetic mechanisms regulating DELLA-mediated signaling pathways also remains paramount for future crop improvement strategies.

Emerging research sheds light on the interplay between genetic mechanisms and epigenetic regulation. Plants possess RNA polymerases IV (Pol IV) and V (Pol V), derived from Pol II, that generate small non-coding RNAs (sncRNAs) and long non-coding RNAs (lncRNAs). These ncRNAs promote transcriptional gene silencing and genome stability. The conserved Mediator complex, which functions with Pol II across eukaryotes, also contributes to non-coding RNA biogenesis in both animals and plants (Kim & Chen, 2011). This complex has been proposed as a key control hub acting in plant adaptive responses to environmental stressful conditions (Chen et al., 2022). Hernández-García et al. (2024) demonstrated that DELLAs function as transcriptional activators through their interactions with the Mediator subunit MED15. This binding is crucial for a subset of DELLA-dependent responses involving co-activation with transcription factors. Moreover, it has been shown that DELLA proteins possess a conserved transcriptional activation domain (TAD) similar to p300/CBP histone acetylases (Thakur et al., 2014; Hernández-García et al.,

2019), thereby potentially linking them to chromatin relaxation and transcription activation via epigenetic pathways.

Boosting Molecular Breeding by CRISPR-Cas TE-mediated Genome Editing

Through precise TE modulation, we can potentially engineer crops with increased yield, robust stress tolerance, and other valuable traits like nutrient efficiency or disease resistance. The CRISPR-Cas gene editing system allows for precise *loci* editing, providing a powerful tool for managing mobile genetic elements in crop genomes (Vaschetto, 2018). This system consists of a guide RNA (gRNA) and a Cas9 enzyme. The gRNA, a short RNA molecule, directs the Cas9 enzyme to a specific DNA sequence. Once at the target site, Cas9 acts as molecular scissors, cutting the DNA at the desired location. This precise cutting allows scientists to remove, add, or alter genetic material, enabling a wide range of applications in biotechnology and medicine.

CRISPR-based epigenome editing is a revolutionary technology that enables precise modification of epigenetic marks, including both DNA methylation and histone modifications (Fig. 6.1). Unlike traditional gene editing techniques, which alters the underlying DNA sequence, CRISPR-based epigenome editing targets the chemical modifications that regulate gene expression without changing the DNA itself. This approach also holds immense potential for identifying, analyzing and manipulating complex crop traits associated with epigenetic rather than genetic variation. CRISPR-based epigenome editing has recently emerged as a transformative strategy for manipulating the activity of specific genomic sequences, including TEs. This innovative epigenetic approach combines the precise targeting capabilities of the deactivated Cas (dCas) protein to bind target sequences and the transient activation of epigenetic remodeling domains to create macromolecular chimeric enzymes capable of directly modifying epigenetic marks, offering a powerful tool for regulating mobility and expression of specific regulatory sequences, without affecting neighboring genes. For instance, certain TEs and repeat sequences known as stress-responsive elements naturally activate and transcribe in response to environmental challenges, contributing to the plant's adaptive responses to surrounding changes. Through targeted epigenetic control, crops can be programmed to activate these stress-responsive sequences and their downstream genes following environmental challenge, thereby pre-priming their defense mechanisms and increasing resilience. Alternatively, precise epigenome modifications can also be harnessed to mitigate the deleterious consequences of TE mobilization under partially fluctuating climate conditions.

The RNAi technology also enables the manipulation of targeted TE and TE-derived sequences within crop genomes. By genetically engineering crop plants to produce targeted siRNAs, researchers can effectively degrade TE-derived

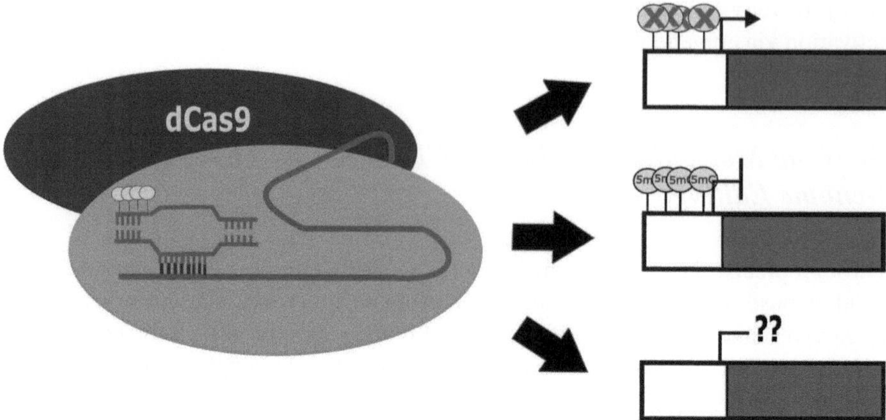

Fig. 6.1 CRISPR-based technologies for epigenome editing. The deactivated Cas enzyme (dCas) without catalytic activity is fused to proteins with epigenetic functions such as DNA methyltransferases and DNA demethylases, creating chimeric constructs capable of activating and deactivating target genes, respectively. The resulting complexes enable efficient and versatile epigenetic editing, making them ideal for modulating gene expression and investigating the roles of distinct genomic sequences. CRISPR-based epigenome editing can be used for silencing the expression of regulatory sequences (Thakur et al., 2014), including potentially harmful transposable elements. Alternatively, this strategy has also been applied to activate genes like *FWA* in *Arabidopsis* (Gallego-Bartolomé et al., 2018). Ligh blue circles indicates methylation marks. Red 'X' marks denote methylation removal in the promoter region of the target gene (brown box: open reading frame, ORF). Question marks indicate potential epigenetic changes (e.g., histone acetylation), which can be accomplished by designing alternative chimeric fusion proteins. dCas 9: purple oval; epigenetic enzyme domain: orange oval; green line: single guide RNA (sgRNA)

transcripts or directly inhibit their translation. This RNAi-based strategy may potentially reduce transposition rates, contributing significantly to genome stability and paving the way for improved crop health.

Conclusion

TEs, not long ago considered selfish parasitic DNA sequences, have emerged as critical partners in host gene regulation. On the other hand, as climate change disrupts weather patterns and stresses ecosystems, traditional crop breeding methods face increasing challenges. Climate change demands crops that can thrive under extreme temperatures, drought, or flooding. In such a scenario, mobile TE sequences harbor an amazing potential for adaptation and crop improvement. By understanding how TE activity influence gene regulation and stress responses, breeders can potentially exploit this major reservoir of genetic and epigenetic variation.

References

Arteaga-Vazquez, M., Sidorenko, L., Rabanal, F. A., Shrivistava, R., Nobuta, K., Green, P. J., et al. (2010). RNA-mediated trans-communication can establish paramutation at the b1 locus in maize. *Proceedings of the National Academy of Sciences, 107*(29), 12986–12991.

Barteri, F. (2019). Impact of transposition on the generation of genetic variability in Prunus crop species.

Blanco-Touriñán, N., Serrano-Mislata, A., & Alabadí, D. (2020). Regulation of DELLA proteins by post-translational modifications. *Plant and Cell Physiology, 61*(11), 1891–1901.

Cahn, J., Regulski, M., Lynn, J., Ernst, E., de Santis Alves, C., Ramakrishnan, S., et al. (2024). MaizeCODE reveals bi-directionally expressed enhancers that harbor molecular signatures of maize domestication. *bioRxiv*. 2024–02.

Casacuberta, J. M., & Santiago, N. (2003). Plant LTR-retrotransposons and MITEs: Control of transposition and impact on the evolution of plant genes and genomes. *Gene, 311*, 1–11.

Chen, J., Yang, S., Fan, B., Zhu, C., & Chen, Z. (2022). The mediator complex: A central coordinator of plant adaptive responses to environmental stresses. *International Journal of Molecular Sciences, 23*(11), 6170.

Chénais, B., Caruso, A., Hiard, S., & Casse, N. (2012). The impact of transposable elements on eukaryotic genomes: From genome size increase to genetic adaptation to stressful environments. *Gene, 509*(1), 7–15.

Clark, R. M., Linton, E., Messing, J., & Doebley, J. F. (2004). Pattern of diversity in the genomic region near the maize domestication gene tb1. *Proceedings of the National Academy of Sciences, 101*(3), 700–707.

Clark, R. M., Wagler, T. N., Quijada, P., & Doebley, J. F. (2006). A distant upstream enhancer at the maize domestication gene tb1 has pleiotropic effects on plant and inflorescence architecture. *Nature Genetics, 38*(5), 594–597.

Doebley, J., Stec, A., & Hubbard, L. (1997). The evolution of apical dominance in maize. *Nature, 386*(6624), 485–488.

Feschotte, C., & Pritham, E. J. (2007). DNA transposons and the evolution of eukaryotic genomes. *Annual Review of Genetics, 41*, 331–368.

Fultz, D., Choudury, S. G., & Slotkin, R. K. (2015). Silencing of active transposable elements in plants. *Current Opinion in Plant Biology, 27*, 67–76.

Galindo-González, L., Sarmiento, F., & Quimbaya, M. A. (2018). Shaping plant adaptability, genome structure and gene expression through transposable element epigenetic control: Focus on methylation. *Agronomy, 8*(9), 180.

Gallego-Bartolomé, J., Gardiner, J., Liu, W., Papikian, A., Ghoshal, B., Kuo, H. Y., et al. (2018). Targeted DNA demethylation of the Arabidopsis genome using the human TET1 catalytic domain. *Proceedings of the National Academy of Sciences, 115*(9), E2125–E2134.

Hedden, P. (2003). The genes of the Green revolution. *Trends in Genetics, 19*(1), 5–9.

Hernández-García, J., Briones-Moreno, A., Dumas, R., & Blázquez, M. A. (2019). Origin of gibberellin-dependent transcriptional regulation by molecular exploitation of a transactivation domain in DELLA proteins. *Molecular Biology and Evolution, 36*(5), 908–918.

Hernández-García, J., Serrano-Mislata, A., Lozano-Quiles, M., Úrbez, C., Nohales, M. A., Blanco-Touriñán, N., et al. (2024). DELLA proteins recruit the mediator complex subunit MED15 to coactivate transcription in land plants. *Proceedings of the National Academy of Sciences, 121*(19), e2319163121.

Hirsch, C. D., & Springer, N. M. (2017). Transposable element influences on gene expression in plants. *Biochimica et Biophysica Acta (BBA)-Gene Regulatory Mechanisms, 1860*(1), 157–165.

Hövel, I., Bader, R., Louwers, M., Haring, M., Peek, K., Gent, J. I., & Stam, M. (2024). RNA-directed DNA methylation mutants reduce histone methylation at the paramutated maize booster1 enhancer. *Plant Physiology, 195*(2), 1161–1179.

Huang, X., Tian, H., Park, J., Oh, D. H., Hu, J., Zentella, R., et al. (2023). The master growth regulator DELLA binding to histone H2A is essential for DELLA-mediated global transcription regulation. *Nature Plants, 9*(8), 1291–1305.

Joly-Lopez, Z., Hoen, D. R., Blanchette, M., & Bureau, T. E. (2016). Phylogenetic and genomic analyses resolve the origin of important plant genes derived from transposable elements. *Molecular Biology and Evolution, 33*(8), 1937–1956.

Kim, Y. J., & Chen, X. (2011). The plant mediator and its role in noncoding RNA production. *Frontiers in Biology, 6,* 125–132.

Kim, W. R., Park, E. G., Lee, Y. J., Bae, W. H., Lee, D. H., & Kim, H. S. (2022). Integration of TE induces cancer specific alternative splicing events. *International Journal of Molecular Sciences, 23*(18), 10918.

Liu, H., Liu, H., Wei, L., Yang, X., & Lin, Z. (2014). A transposable element insertion disturbed starch synthase gene SSIIb in maize. *Molecular Breeding, 34,* 1159–1171.

Louwers, M., Haring, M., & Stam, M. (2005). When alleles meet: Paramutation. In P. Meyer (Ed.), *Annual plant reviews volume 19: Plant Epigenetics* (pp. 134–173). Blackwell.

Makarevitch, I., Waters, A. J., West, P. T., Stitzer, M., Hirsch, C. N., Ross-Ibarra, J., & Springer, N. M. (2015). Transposable elements contribute to activation of maize genes in response to abiotic stress. *PLoS Genetics, 11*(1), e1004915.

Martienssen, R. A. (2003). Maintenance of heterochromatin by RNA interference of tandem repeats. *Nature Genetics, 35*(3), 213–214.

Pathak, R. U., Phanindhar, K., & Mishra, R. K. (2024). Transposable elements as scaffold/matrix attachment regions: Shaping organization and functions in genomes. *Frontiers in Molecular Biosciences, 10,* 1326933.

Patty, B. J., & Hainer, S. J. (2020). Non-coding RNAs and nucleosome remodeling complexes: An intricate regulatory relationship. *Biology, 9*(8), 213.

Pegler, J. L., Oultram, J. M., Mann, C. W., Carroll, B. J., Grof, C. P., & Eamens, A. L. (2023). Miniature inverted-repeat transposable elements: Small DNA transposons that have contributed to plant MICRORNA gene evolution. *Plants, 12*(5), 1101.

Prakash, N. R., Chhabra, R., Zunjare, R. U., Muthusamy, V., & Hossain, F. (2020). Molecular characterization of teosinte branched1 gene governing branching architecture in cultivated maize and wild relatives. *3 Biotech, 10*(2), 77.

Sarnowska, E. A., Rolicka, A. T., Bucior, E., Cwiek, P., Tohge, T., Fernie, A. R., et al. (2013). DELLA-interacting SWI3C core subunit of switch/sucrose nonfermenting chromatin remodeling complex modulates gibberellin responses and hormonal cross talk in Arabidopsis. *Plant Physiology, 163*(1), 305–317.

Schrader, L., & Schmitz, J. (2019). The impact of transposable elements in adaptive evolution. *Molecular Ecology, 28*(6), 1537–1549.

Sidorenko, L., Dorweiler, J. E., Cigan, A. M., Arteaga-Vazquez, M., Vyas, M., Kermicle, J., et al. (2009). A dominant mutation in mediator of paramutation 2, one of three second-largest subunits of a plant-specific RNA polymerase, disrupts multiple siRNA silencing processes. *PLoS Genetics, 5*(11), e1000725.

Studer, A. J., & Doebley, J. F. (2012). Evidence for a natural allelic series at the maize domestication locus teosinte branched1. *Genetics, 191*(3), 951–958.

Studer, A., Zhao, Q., Ross-Ibarra, J., & Doebley, J. F. (2011). Identification of a functional transposon insertion in the maize domestication gene tb1. *Nature Genetics, 43*(11), 1160–1163.

Thakur, J. K., Yadav, A., & Yadav, G. (2014). Molecular recognition by the KIX domain and its role in gene regulation. *Nucleic Acids Research, 42*(4), 2112–2125.

Thieme, M., & Bucher, E. (2018). Transposable elements as tool for crop improvement. In *Advances in botanical research* (Vol. 88, pp. 165–202). Academic.

Tikhonov, A. P., Bennetzen, J. L., & Avramova, Z. V. (2000). Structural domains and matrix attachment regions along colinear chromosomal segments of maize and sorghum. *The Plant Cell, 12*(2), 249–264.

Ulukapi, K., & Nasircilar, A. G. (2018). Induced mutation: Creating genetic diversity in plants. In *Genetic diversity in plant species-characterization and conservation*. IntechOpen.

Vaschetto, L. M. (2015). Exploring an emerging issue: Crop epigenetics. *Plant Molecular Biology Reporter, 33*, 751–755.

Vaschetto, L. M. (2016). Miniature inverted-repeat transposable elements (MITEs) and their effects on the regulation of major genes in cereal grass genomes. *Molecular Breeding, 36*(3), 30.

Vaschetto, L. M. (2018). Modulating signaling networks by CRISPR/Cas9-mediated transposable element insertion. *Current Genetics, 64*(2), 405–412.

Vicient, C. M. (2010). Transcriptional activity of transposable elements in maize. *BMC Genomics, 11*, 1–10.

Wang, L., Gao, J., Wang, C., Xu, Y., Li, X., Yang, J., et al. (2022). Comprehensive analysis of long non-coding RNA modulates axillary bud development in tobacco (Nicotiana tabacum l.). *Frontiers in Plant Science, 13*, 809435.

Wilhelm, E. P., Howells, R. M., Al-Kaff, N., et al. (2013). Genetic characterization and mapping of the Rht-1 homoeologs and flanking sequences in wheat. *Theoretical and Applied Genetics, 126*, 1321–1336. https://doi.org/10.1007/s00122-013-2055-3

Zhou, L., Zhang, J., Yan, J., & Song, R. (2011). Two transposable element insertions are causative mutations for the major domestication gene teosinte branched1 in modern maize. *Cell Research, 21*(8), 1267–1270.

Zhou, M., Hu, H., Liu, Z., & Tang, D. (2016). Two active bamboo mariner-like transposable elements (Ppmar1 and Ppmar2) identified as the transposon-based genetic tools for mutagenesis. *Molecular Breeding, 36*(12), 163.

Chapter 7
Plant Epitranscriptomics

Beyond the DNA sequence itself, the regulation of gene expression involves a complex interplay between various mechanisms. Epitranscriptomics, a rapidly evolving field, sheds light on the dynamic and reversible modifications occurring on messenger RNA (mRNA) molecules. These modifications, often mediated by enzymatic processes, introduce chemical alterations to RNA bases, influencing mRNA stability, structure, and ultimately, its interaction with the translational machinery. Deciphering the plant epitranscriptomic code and its role in cellular processes holds immense potential for understanding gene regulation patterns and enhancing agronomic traits.

The RNA World: A History on Writers, Erasers, and Readers

RNA modifiers can be defined as a set of proteins or protein complexes that act on RNA molecules to regulate various cellular processes. RNA writers play a crucial role in regulating gene expression by chemically modifying RNA molecules. An example of this are RNA methyltransferases, which can add methyl groups (–CH3) to mRNA, influencing its stability and translation efficiency. Conversely, RNA erasers remove chemical marks, resetting the epigenetic state and potentially reversing the effects (Shinde et al., 2023). Moreover, RNA readers are proteins that bind to the chemically modified RNA (e.g., by methylated nucleotides), interpreting these marks and mediating downstream effects.

The RNA epitranscriptomic code constitutes a fundamental regulatory layer that orchestrates intricate cellular processes, serving as a pivotal determinant of plant development. For instance, N6-methyladenosine (m6A) deposition by RNA writers plays a key role in plant stress responses, regulating gene expression under adverse conditions (Hu et al., 2022; Reichel et al., 2019). In *Arabidopsis*, specific proteins involved in m6A modifications have been identified, and their dysregulation leads

L. M. Vaschetto, *Epigenetics in Crop Improvement*, https://doi.org/10.1007/978-3-031-73176-1_7

to developmental abnormalities and altered timing of reproductive development (Shen et al., 2019). Nevertheless, these enzymes don't operate in isolation. They often serve as integral components in stress responses. In conditions of stress, enzymatic writers are activated to modify RNA molecules participating in critical response pathways, such as drought tolerance, heat shock response, or pathogen defense. Thus, RNA modifications can efficiently fine-tune gene expression, enabling plants to face adverse environmental conditions.

RNA erasers also play crucial roles in plant development, organelle biogenesis, environmental adaptation, and signaling (Shinde et al., 2023). Their discovery has shed light on the intricate plant epitranscriptomic landscape, where RNA modifications dynamically regulate gene expression (Shen et al., 2019). For instance, the RNA eraser *ALKBH10B* specifically recognizes and demethylates m6A on RNA molecules, reinstating the adenosine residue to its unmodified state (Duan et al., 2017). This epitranscriptomic event can alter RNA stability, splicing, or translation, ultimately impacting various downstream processes.

RNA readers, proteins equipped with specialized RNA-binding domains, play a crucial role in deciphering the epitranscriptomic code. As a result, readers have essential roles in gene regulation and developmental pathways. These proteins can interpret and bind to specific chemical groups on the RNA molecule, influence chromatin remodeling and ultimately affect transcriptional events. Some RNA readers have been shown to mediate defense responses against viral infection (Shinde et al., 2023). In plants, there are several types of RNA readers, including Methyl-Binding Domain proteins (MBD), SET and RING finger-associated (SRA) domain-containing proteins, and UHRF1 homologs (Grimanelli & Ingouff, 2020). RNA reader domains recognize and interact with distinct RNA modifications, thus bridging the gap between RNA modifications and downstream cellular events. Their interactions with modified transcripts trigger a cascade of signaling pathways, leading to diverse functional outcomes. The impact of reader-modification interactions is highly context-dependent, varying depending on the specific reader and its cognate modification. The recognition of RNA modifications by readers represents a critical control point in plant gene expression. These interactions trigger various molecular mechanisms, such as the recruitment of effector proteins, alterations in the RNA secondary structure, and changes in RNA-protein interactions, ultimately determining the functional consequences of modification recognition and shaping cellular responses.

A Overview of RNA Modifications in Plants

Over the last years, the rapidly expanding field of plant epitranscriptomics have shed new light into a previously unknown layer of gene regulation. This type of gene regulation involves a diverse array of RNA modifications, including but not limited to, N6-methyladenosine (m6A), 5-methylcytosine (m5C), pseudouridine (Ψ), 2'-O-methylation (Nm), 8-hydroxyguanine (8-OHG), and 8-nitroguanine

(8-NG). These RNA modifications have been identified in various types of RNAs where they are associated with multiple metabolic pathways such as those involved in stress response, RNA stability, translation, and transport (Hu et al., 2022; Shoaib et al., 2022; Xie et al., 2023). Briefly, I will review some of the most well studied epitranscriptomic modifications and their functional significance in plant development and stress responses.

Pseudouridylation

Isomerization of uridine generates pseudouridine (Ψ), a ribonucleotide modification with an additional H-bond donor capable of enhancing RNA stability. It is an abundant RNA modification found in various RNA types and plays important roles in distinct biological processes (Netzband & Pager, 2020). The structural alterations induced by pseudouridine are indispensable for the functionality of specific RNA molecules. In both plants and animals, pseudouridine marks small RNAs, enabling their transport into the nucleus, where they can influence gene expression. In *Arabidopsis*, for instance, pseudouridine modifications have been recently detected in both small non-coding RNAs and their precursors. Rowan et al. (2023) observed significant enrichment of pseudouridine in germline small RNAs, particularly in epigenetically activated siRNAs (easiRNAs) within *Arabidopsis* pollen. This finding suggests a conserved role for pseudouridine in marking epigenetically inherited small RNAs within the plant germline.

Enzymes involved in pseudouridylation have shown to be crucial for normal plant development. Also in *Arabidopsis*, Niu et al. (2022) observed that mutants lacking the *FCS1* gene (*fcs1-1*), which encodes a pseudouridine synthase, exhibit delayed development, reduced fertility, and significant size reduction compared to wild-type plants. The researchers attributed these phenotypes to a substantial decrease in mitochondrial 26S rRNA pseudouridylation, likely disrupting mitochondrial translation in *fcs1-1* mutants. In rice, Wang, Sun, et al. (2022a) identified a role for chloroplast ribosomal RNA (rRNA) pseudouridylation cold tolerance at the seedling stage, suggesting potential applications in crop improvement. OsPUS1, another pseudouridine synthase, accumulates during cold stress and binds to chloroplast precursor rRNAs (pre-rRNAs) to catalyze pseudouridylation. These modifications appear essential for rRNA processing, as evidenced by the reduction of mature chloroplast rRNAs and accumulation of pre-rRNAs in the *ospus1-1* mutant under cold conditions. Xie et al. (2022) reported tissue-specific and stress-responsive expression patterns of maize pseudouridine synthases. Remarkably, these enzymes exhibited differential expression under heat and salt stress conditions, suggesting potential roles in stress response pathways. Furthermore, subcellular localization analysis revealed that individual pseudouridine synthase subfamilies localize to distinct organelles, including the nucleus, cytoplasm, and chloroplasts (Xie et al., 2022).

N6-methyladenosine

N6-methyladenosine (m6A) involves the methylation of the N6 position on adenosine residues. This prevalent RNA modification occurs in diverse RNA species, including messenger RNA (mRNA), transfer RNA (tRNA), ribosomal RNA (rRNA), and various types of ncRNAs. Plant m6A modifications play critical roles in post-transcriptional gene regulation by impacting mRNA stability, splicing patterns, and translation efficiency. These roles range from facilitating the formation of membraneless biomolecular condensates through liquid-liquid phase separation (LLPS), crucial for various cellular processes (Kang & Xu, 2023), to maintaining photosynthetic efficiency under light stress conditions (Zhang et al., 2022). The formation of membraneless biomolecular condensates is essential for various cellular processes, including embryogenesis, floral transition, photosynthesis, pathogen defense, and stress responses (Kang & Xu, 2023).

m6A modifications on transcripts of salt-resistance genes may also be crucial determinants of crop salt tolerance. Zheng et al. (2021) observed significant alterations in m6A patterns within sweet sorghum (*Sorghum bicolor*) under salt stress. The salt-sensitive 'Roma' sorghum lines exhibit increased m6A modification on mRNAs associated with salt-resistance genes. This suggests that m6A modifications may enhance mRNA stability, potentially contributing to improved stress tolerance (Zheng et al., 2021). N6-methyladenosine (m6A) modifications also play a critical role in plant immune responses to viruses. Martínez-Pérez et al. (2023) demonstrated that the infectivity of the alfalfa mosaic virus (AMV), a positive-strand RNA virus, relies on demethylation of its viral RNA by a cellular m6A demethylase, ALKBH9B. In this regard, plant YTHDF proteins act as direct effectors of antiviral immunity by recognizing m6A-containing RNA viruses (Martínez-Pérez et al., 2023). Zhang et al. (2021) reported that m6A methylation in virus-infected rice plants primarily associates with genes exhibiting low expression. Interestingly, they observed increased m6A levels in response to viral infection, including genes involved in antiviral pathways (i.e., RNA silencing, resistance, and antiviral phytohormone metabolism). This suggests a potential link between m6A methylation and the regulation of antiviral gene expression (Zhang et al., 2021). In barley, Rudy et al. (2023) explored the role of m6A modifications as a potential metabolic switch between plant cell survival and death during leaf senescence. They observed distinct patterns of m6A modification undergoing dark-induced senescence compared to developmental senescence. Consequently, the authors of this study interestingly suggested the potential existence of an additional, yet unidentified epigenetic switch regulating the balance between cell survival and death (Rudy et al., 2023).

5-methylcytosine (m5C)

5-methylcytosine (m5C) is another prevalent RNA modification. It is characterized by methylation at the C5 position of cytosine residues. The modification m5C is an epitranscriptomic mark that profoundly influences gene expression, impacting RNA stability, translation, transcription, nuclear export, and cleavage across diverse metabolic and cellular processes. Recent studies highlight the importance of m5C in transcription. Yang et al. (2019) reported a significant enrichment of m5C in *Arabidopsis* mobile transcripts, suggesting a potential regulatory role in graft-transmissible RNA movement and its downstream effects. Also in *Arabidopsis*, Zhang et al. (2023) suggested a connection between RNA m5C modification and the histone modification H3K27me3. The authors of this study observed that m5C marks are primarily enriched in genomic regions devoid of H3K27me3, observed in both wild-type and *emf1* mutant plants. EMF1 is a plant-specific Polycomb group (PcG) protein with epigenetic regulatory roles. The *emf1* mutant plants exhibit increased m5C levels on target genes, leading to a decrease in target mRNA transcripts linked to photosynthesis and chloroplast genes. Consequently, these findings suggest there is an interplay between m5C and histone modifications, potentially regulating gene expression.

m5C marks have been shown to be particularly important in the context of climate change. Tang et al. (2020) identified OsNSUN2, an RNA 5-methylcytosine (m5C) methyltransferase in rice, that enhances adaptation to high temperature. They demonstrated that OsNSUN2 mediates m5C modification of mRNAs associated with photosynthesis and detoxification systems. Heat stress increased OsNSUN2-dependent m5C marks, suggesting a role in the heat response. Remarkably, *osnsun2* mutants exhibited compromised photosystems under high temperatures, failing to repair under tolerable heat stress. This resulted in reduced photosynthetic efficiency and accumulation of reactive oxygen species, thereby highlighting the importance of OsNSUN2-mediated m5C modification for thermotolerance (Tang et al., 2020).

RNA Uridylation

RNA uridylation is a well-known modification that regulates gene expression post-transcriptionally in eukaryotes,. This process involves the addition of uridine residues to the 3′ end of RNA molecules by terminal uridylyl transferases, a sub-group of non-canonical nucleotidyl transferases. In *Arabidopsis*, HEN1 SUPPRESSOR 1 (HESO1) and UTP:RNA URIDYLYLTRANSFERASE (URT1) are the best-characterized enzymes known to catalyze RNA uridylation (Wang, Kong, et al., 2022b). RNA uridylation has been shown to be a key post-transcriptional modification shaping small ncRNA landscapes in *Arabidopsis*, where HESO1 and URT1 cooperatively uridylate AGO-bound unmethylated miRNAs and siRNAs, targeting them for degradation (Wang, Kong, et al., 2022b). Overexpression of HESO1 led to morphological defects and reduced miRNA accumulation, suggesting that uridylation may destabilize unmethylated miRNAs (Ren et al., 2012). In 2015, Wang et al. identified URT1 as a functional paralog of the *Arabidopsis* RNA uridylyltransferase HESO1. URT1 preferentially interacts with AGO1 and assumes a dominant role in miRNA uridylation during HESO1 deficiency. Interestingly, URT1 disruption does not appear to affect heterochromatic siRNA uridylation, suggesting the presence of additional enzymes mediating uridylation within the siRNA pathway (Wang et al., 2015).

Furthermore, HESO1 and URT1 also target mRNA transcripts for degradation via uridylation. Zuber et al. (2018) investigated the cooperative action of URT1 and HESO1 in uridylating 5′ fragments derived from RISC-cleaved mRNAs. They analyzed tailing and trimming patterns of RISC-cleaved MYB33 and SPL13 transcripts in wild-type and mutant *Arabidopsis* lines for HESO1 and URT1. Remarkably, it was observed that the absence of uridylation led to the accumulation of 5′ fragments exhibiting short deletions (nibbling) near the RISC-cleavage site (Zuber et al., 2018). Zuber et al. previously identified a mechanism by which URT1 regulates the extent of mRNA deadenylation, a key process for mRNA decay. While URT1-independent uridylation exists, only URT1 activity prevents excessive deadenylation, highlighting its specific function in maintaining mRNA stability (Zuber et al., 2016).

Functional Significance of RNA Modifications

Beyond the canonical genetic code, the RNA epitranscriptome emerges as a crucial layer of gene regulation, influencing various cellular processes across the eukaryotic kingdom. For instance, RNA modifications control translation efficiency, ultimately modulating stress pathways. In *Arabidopsis*, Wang et al. (2023) reduced the levels of the mRNA adenosine methylase A (MTA) via RNAi and demonstrated that cold-induced m6A depletion, particularly in 3′ UTRs, plays a crucial role in modulating translation, ultimately altering chilling response. Moreover, m6A marks within the coding region or 5′ UTR can also enhance translation efficiency by protecting it from degradation pathways (Meyer et al., 2015). The delicate equilibrium among distinct types of epitranscriptomic marks likely dictates the destination of individual RNA sequences, either facilitating their retention or targeting them for degradation in response to cellular needs. Consequently, RNA modifications can exert a profound influence on overall gene expression levels. The impact of these marks on RNA further extends to translation efficiency, effectively controlling the rate at which an encoded protein is synthesized. This dynamic regulation allows plants to adapt to fluctuating environmental conditions by prioritizing specific mRNAs when necessary.

RNA marks also exert a profound influence on the tridimensional properties of RNA molecules. By introducing structural perturbations, such as altered base-pairing patterns or steric clashes, pseudouridine can reshape secondary and tertiary structures (Zhao & He, 2015), leading to downstream functional consequences. These modifications also enhance the stability of RNA structures. For instance, RNA marks may potentially modulate tRNA folding patterns (Ramos & Fu, 2019), guaranteeing their fidelity during translation. Moreover, pseudouridine is present in functionally critical regions of ribosomal RNAs (Kierzek et al., 2014), thereby this RNA modification might enhance the structural integrity of ribosomal RNAs (rRNAs) and optimize their function.

Exploiting the Epitranscriptomic Code in Crop Improvement

Epitranscriptomic marks have recently emerged as master regulators of RNA function and molecular landscapes, dynamically shaping the structure and behavior of diverse RNA species, including mRNAs, tRNAs, rRNAs, and non-coding RNAs. These RNA modifications represent intricate biochemical processes that can potentially orchestrate gene regulation, developmental programs, and stress responses, ultimately enabling plants to adapt to challenging environmental conditions. Nonetheless, despite the identification of more than 150 types of RNA marks across the tree of life, harnessing this knowledge to optimize the epitranscriptomic code for enhanced crop productivity has not yet been successful (Ramakrishnan et al., 2022). Realizing the potential of RNA modifications in crop improvement hinges on their ability to influence crucial agronomic traits.

Gene regulation networks governed by epitranscriptomic modifications offers an unique opportunity to fine-tune crop traits, including their resilience in a climate change scenario. Further research on RNA writers, erasers, and readers is necessary to unlock their full potential. One of the most promising avenues for harnessing RNA marks in crop improvement is found in the emerging field of epitranscriptomic engineering. This approach encompasses strategies for precisely and accurately manipulating RNA modifications. Targeted epitranscriptomic editing involves the precise addition or removal of specific RNA modifications at desired *loci* within the genome (Xia et al., 2021). Shi et al. (2024) developed a CRISPR-Cas13a-based platform to precisely manipulate N6-methyladenosine (m6A) modifications in plant mRNAs. By fusing m6A writers and erasers to a catalytically inactive Cas13a, they achieved targeted m6A deposition and removal in both *Nicotiana benthamiana* and *Arabidopsis thaliana*. Functional validation through *SHR* mRNA editing demonstrated that m6A modifications can regulate gene expression and influence plant growth. Manipulating RNA writers and readers thus presents an interesting avenue that deserves investigation. These proteins can be manipulated to shape the manner that the epitranscriptomic code is interpreted within the cellular context. By engineering these proteins, researchers can influence the deposition and removal of RNA modifications, orchestrating signaling pathways to achieve desired crop traits.

Conclusions

By influencing RNA stability, translation, and localization, the epitranscriptomic code offer multiple dynamic layers of gene regulation beyond the DNA sequence itself. Molecular breeding efforts can take advantage of this knowledge by identifying and selecting plants with epitranscriptomic profiles that enhance tolerance to climate change stresses such as drought, heat, and salinity. Targeting specific RNA marks or the enzymes responsible for adding/removing/reading them could lead to the development of crop varieties with improved resilience. As research on epitranscriptomic marks continues to advance, their potential for revolutionizing molecular breeding becomes increasingly more evident.

References

Duan, H. C., Wei, L. H., Zhang, C., Wang, Y., Chen, L., Lu, Z., et al. (2017). ALKBH10B is an RNA N 6-methyladenosine demethylase affecting Arabidopsis floral transition. *The Plant Cell, 29*(12), 2995–3011.

Grimanelli, D., & Ingouff, M. (2020). DNA methylation readers in plants. *Journal of Molecular Biology, 432*(6), 1706–1717.

Hu, J., Cai, J., Xu, T., & Kang, H. (2022). Epitranscriptomic mRNA modifications governing plant stress responses: Underlying mechanism and potential application. *Plant Biotechnology Journal, 20*(12), 2245–2257. https://doi.org/10.1111/pbi.13913

Kang, H., & Xu, T. (2023). N6-methyladenosine RNA methylation modulates liquid–liquid phase separation in plants. *The Plant Cell, koad103*, 3205–3213.

Kierzek, E., Malgowska, M., Lisowiec, J., Turner, D. H., Gdaniec, Z., & Kierzek, R. (2014). The contribution of pseudouridine to stabilities and structure of RNAs. *Nucleic Acids Research, 42*(5), 3492–3501.

Martínez-Pérez, M., Aparicio, F., Arribas-Hernández, L., Tankmar, M. D., Rennie, S., von Bülow, S., et al. (2023). Plant YTHDF proteins are direct effectors of antiviral immunity against an N6-methyladenosine-containing RNA virus. *The EMBO Journal, 42*(18), e113378.

Meyer, K. D., Patil, D. P., Zhou, J., Zinoviev, A., Skabkin, M. A., Elemento, O., et al. (2015). 5′ UTR m6A promotes cap-independent translation. *Cell, 163*(4), 999–1010.

Netzband, R., & Pager, C. T. (2020). Epitranscriptomic marks: Emerging modulators of RNA virus gene expression. *Wiley Interdisciplinary Reviews: RNA, 11*(3), e1576.

Niu, Y., Zheng, Y., Zhu, H., Zhao, H., Nie, K., Wang, X., et al. (2022). The Arabidopsis mitochondrial pseudouridine synthase homolog FCS1 plays critical roles in plant development. *Plant and Cell Physiology, 63*(7), 955–966.

Ramakrishnan, M., Rajan, K. S., Mullasseri, S., Palakkal, S., Kalpana, K., Sharma, A., et al. (2022). The plant epitranscriptome: Revisiting pseudouridine and 2′-O-methyl RNA modifications. *Plant Biotechnology Journal, 20*(7), 1241–1256.

Ramos, J., & Fu, D. (2019). The emerging impact of tRNA modifications in the brain and nervous system. *Biochimica et Biophysica Acta (BBA)-Gene Regulatory Mechanisms, 1862*(3), 412–428.

Reichel, M., Köster, T., & Staiger, D. (2019). Marking RNA: m6A writers, readers, and functions in Arabidopsis. *Journal of Molecular Cell Biology, 11*(10), 899–910.

Ren, G., Chen, X., & Yu, B. (2012). Uridylation of miRNAs by hen1 suppressor1 in Arabidopsis. *Current Biology, 22*, 695–700.

Rowan, P. H., Jakub, D., Valentina, M., de Santis Alves, C., Filipe, B., Van Ex, F., Ann, L., Mateusz, B., Tommaso, L., Hendrick, A. G., Schorn, A. J., Kouzarides, T., & Martienssen, R. A. (2023). Pseudouridine guides germline small RNA transport and epigenetic inheritance. *Nature Structural & Molecular Biology*. https://doi.org/10.1101/2023.05.27.542553

Rudy, E., Grabsztunowicz, M., Arasimowicz-Jelonek, M., Tanwar, U. K., Maciorowska, J., & Sobieszczuk-Nowicka, E. (2023). N6-methyladenosine (m6A) RNA modification as a metabolic switch between plant cell survival and death in leaf senescence. *Frontiers in Plant Science, 13*, 1064131.

Shen, L., et al. (2019). Messenger RNA modifications in plants. *Trends in Plant Science, 24*(4), 328–341.

Shi, C., Zou, W., Liu, X., Zhang, H., Li, X., Fu, G., et al. (2024). Programmable RNA N6-methyladenosine editing with CRISPR/dCas13a in plants. *Plant Biotechnology Journal., 22*, 1867–1880.

Shinde, H., Dudhate, A., Kadam, U. S., & Hong, J. C. (2023). RNA methylation in plants: An overview. *Frontiers in Plant Science, 14*, 1132959.

Shoaib, Y., Usman, B., Kang, H., & Jung, K. H. (2022). Epitranscriptomics: An additional regulatory layer in plants' development and stress response. *Plants, 11*(8), 1033.

Tang, Y., Gao, C. C., Gao, Y., Yang, Y., Shi, B., Yu, J. L., et al. (2020). OsNSUN2-mediated 5-methylcytosine mRNA modification enhances rice adaptation to high temperature. *Developmental Cell, 53*(3), 272–286.

Wang, X., Zhang, S., Dou, Y., Zhang, C., Chen, X., Yu, B., & Ren, G. (2015). Synergistic and independent actions of multiple terminal nucleotidyl transferases in the 3'tailing of small RNAs in Arabidopsis. *PLoS Genetics, 11*(4), e1005091.

Wang, Z., Sun, J., Zu, X., Gong, J., Deng, H., Hang, R., et al. (2022a). Pseudouridylation of chloroplast ribosomal RNA contributes to low temperature acclimation in rice. *New Phytologist, 236*(5), 1708–1720.

Wang, X., Kong, W., Wang, Y., Wang, J., Zhong, L., Lao, K., et al. (2022b). Uridylation and the SKI complex orchestrate the Calvin cycle of photosynthesis through RNA surveillance of TKL1 in Arabidopsis. *Proceedings of the National Academy of Sciences, 119*(38), e2205842119.

Wang, S., Wang, H., Xu, Z., Jiang, S., Shi, Y., Xie, H., et al. (2023). m6A mRNA modification promotes chilling tolerance and modulates gene translation efficiency in Arabidopsis. *Plant Physiology, 192*(2), 1466–1482.

Xia, Z., Tang, M., Ma, J., Zhang, H., Gimple, R. C., Prager, B. C., et al. (2021). Epitranscriptomic editing of the RNA N6-methyladenosine modification by dCasRx conjugated methyltransferase and demethylase. *Nucleic Acids Research, 49*(13), 7361–7374.

Xie, Y., Gu, Y., Shi, G., He, J., Hu, W., & Zhang, Z. (2022). Genome-wide identification and expression analysis of pseudouridine synthase family in Arabidopsis and maize. *International Journal of Molecular Sciences, 23*(5), 2680.

Xie, Y., Chan, L. Y., Cheung, M. Y., Li, M. W., & Lam, H. M. (2023). Current technical advancements in plant epitranscriptomic studies. *The Plant Genome, e20316*.

Yang, L., Perrera, V., Saplaoura, E., Apelt, F., Bahin, M., Kramdi, A., et al. (2019). m5C methylation guides systemic transport of messenger RNA over graft junctions in plants. *Current Biology, 29*(15), 2465–2476.

Zhang, K., Zhuang, X., Dong, Z., Xu, K., Chen, X., Liu, F., & He, Z. (2021). The dynamics of N 6-methyladenine RNA modification in interactions between rice and plant viruses. *Genome Biology, 22*(1), 189.

Zhang, M., Zeng, Y., Peng, R., Dong, J., Lan, Y., Duan, S., et al. (2022). N6-methyladenosine RNA modification regulates photosynthesis during photodamage in plants. *Nature Communications, 13*(1), 7441.

Zhang, D., Guo, W., Wang, T., Wang, Y., Le, L., Xu, F., et al. (2023). RNA 5-methylcytosine modification regulates vegetative development associated with H3K27 trimethylation in Arabidopsis. *Advanced Science, 10*(1), 2204885.

Zhao, B. S., & He, C. (2015). Pseudouridine in a new era of RNA modifications. *Cell Research, 25*(2), 153–154.

Zheng, H., Sun, X., Li, J., Song, Y., Song, J., Wang, F., et al. (2021). Analysis of N6-methyladenosine reveals a new important mechanism regulating the salt tolerance of sweet sorghum. *Plant Science, 304*, 110801.

Zuber, H., Scheer, H., Ferrier, E., Sement, F. M., Mercier, P., Stupfler, B., & Gagliardi, D. (2016). Uridylation and PABP cooperate to repair mRNA deadenylated ends in Arabidopsis. *Cell Reports, 14*(11), 2707–2717.

Zuber, H., Scheer, H., Joly, A. C., & Gagliardi, D. (2018). Respective contributions of URT1 and HESO1 to the uridylation of 5′ fragments produced from RISC-cleaved mRNAs. *Frontiers in Plant Science, 9*, 402087.

Chapter 8
Seed Dormancy and Germination as Models for Understanding Epigenetic Programming

Seed dormancy, an adaptation maximizing seedling survival and reproductive success, fine-tunes germination timing. Germination, the transition from dormancy to active growth, also requires intricate regulation for adaptation and survival. Epigenetic modifications act as dynamic molecular switches, influencing gene expression throughout these crucial stages. The dynamic epigenetic landscape allows plants to integrate environmental cues, facilitating precise spatiotemporal control of seed emergence. Consequently, seeds can optimize timing and location for successful establishment and reproduction. Unraveling the epigenetic mechanisms governing complex adaptive traits like seed dormancy and germination holds immense promise for manipulating physiological responses and adapting plants to challenging environmental conditions.

Environmental Factors Affecting Seed Dormancy and Germination

Environmental stresses significantly influence seed dormancy and germination dynamics at both phenotypic and genetic levels, shaping cellular processes for optimal seedling survival. For example, desert plants may exhibit adaptations requiring prolonged dryness for germination, while some forest species rely on specific photoperiods to break dormancy. Similarly, environmental cues trigger or inhibit germination through epigenetic mechanisms. Temperature fluctuations can significantly impact epigenetic programming during seed dormancy and germination, leading to alterations in DNA methylation and histone modifications, ultimately affecting gene expression. Cold stress, a common dormancy inducer, can trigger increased DNA methylation at *loci* associated with germination genes, effectively silencing them. Zhang et al. (2012a) observed enhanced global DNA methylation in young strawberry leaves upon dormancy induction by environmental factors, thereby suggesting

L. M. Vaschetto, *Epigenetics in Crop Improvement*,
https://doi.org/10.1007/978-3-031-73176-1_8

that external conditions can influence DNA methylation patterns to regulate seed dormancy. Gomez-Cabellos et al. (2022) proposed a link between the level of seed dormancy and epigenetic modifications in *Capsella bursa-pastoris*. They hypothesized that global DNA methylation levels increase, while histone H4 acetylation (H4Ac) marks decrease, as seed dormancy deepens over time.

For some plant species, light is a critical environmental signal that breaks dormancy and initiates seed germination. In *Arabidopsis*, Gu et al. (2019) identified SUVH5, a histone H3 lysine 9 methyltransferase, as a positive regulator of light-mediated seed germination. Loss of function in SUVH5 led to decreased phytochrome B (PHYB)-dependent germination, highlighting the role of histone methylation in this process. Beyond methylation, histone modifications play a broader role in early plant development. Benhamed et al. (2006) observed that the *Arabidopsis* histone acetyltransferases GCN5 and TAF1/HAF2, along with the histone deacetylase HD1/HDA19, are crucial for light regulation of growth and gene expression, thereby evidencing a critical role for histone acetylation in this response. Humidity and water availability also influence seed dormancy and germination through epigenetic regulation. Malabarba et al. (2021) revealed that DNA demethylation by ROS1 impairs germination by altering expression of germination-related genes in *Arabidopsis*. They observed increased differentially methylated regions (DMRs) in promoters and body regions of these genes under severe stress conditions, highlighting the role of DNA methylation dynamics in response to water availability.

Epigenetic mechanisms also regulate seed dormancy and germination through hormonal pathways. In *Arabidopsis*, Zhou et al. (2005) showed that HDA6 inhibits the transcription of target genes involved in both phytohormone signaling pathways, suggesting a complex interplay between epigenetics, hormone signaling, and seed dormancy/germination. Moreover, it has been shown that histone deacetylases HDA6 and HDA19 repress embryo-specific transcription factors that regulate seed maturation, including *LEAFY COTYLEDON1* (*LEC1*), *LEAFY COTYLEDON2* (*LEC2*), *FUSCA3* (*FUS3*), and *ABSCISIC ACID INSENSITIVE3* (*ABI3*) by reducing H3K9 and H3K4 acetylation levels during post germinative stages (Zhou et al., 2013). Interestingly, it has also been demonstrated that ethylene and jasmonic acid induce *HDA6* and *HDA19* expression (Luján-Soto & Dinkova, 2021).

Revisiting Epigenetic Modifications During Dormancy and Germination

While global DNA methylation levels remain relatively stable during dormancy, they undergo dynamic alterations during germination, with a decrease often associated with the activation of genes essential for growth (Narváez et al., 2022). Thus, DNA demethylation acts as a crucial regulator of seed germination, particularly under abiotic stress conditions. Malabarba et al. (2021) suggested that heat stress during seed development leads to changes in gene expression and DNA methylation, thereby altering seed germination and longevity. Chilling exposure and subsequent dormancy induction are potentially linked to DNA methylation. Rothkegel et al. (2017) revealed that DNA methylation directly targets MADS-box genes, key regulators in plant development, during the epigenetic control of seed dormancy in sweet cherry (*Prunus avium* L.). DNA methylation likely regulates seed dormancy by modulating transcription factor access to DNA. Methylated regions become less accessible, silencing gene expression during dormancy. Conversely, favorable conditions trigger active demethylation, reactivating germination-related genes. This dynamic balance between methylation and demethylation can potentially orchestrate the timing and success of seed germination.

Seed dormancy depends on a complex interplay between histone modifications. Specific modifications likely activate dormancy genes while repressing germination-associated genes. Several histone modifiers fine-tune this balance. In *Arabidopsis*, LDL1 and LDL2, histone demethylases, act as dormancy suppressors by promoting dormancy gene expression (Zhao et al., 2015). Conversely, histone deacetylases HD2A and HD2B promote dormancy by silencing these genes (Han et al., 2023), highlighting the contrasting roles of different histone modifications in regulating dormancy and germination. The interplay between histone modifiers extends beyond individual enzymes. Zhou et al. (2020) revealed a cooperative role for the histone methyltransferase SUVH5 and the deacetylase HDA19 in repressing seed dormancy. Additionally, histone chaperones like HIRA influence chromatin dynamics during seed dormancy. Layat et al. (2021) observed that mutations in *HIRA*, which mediates the deposition of the H3.3 histone variant in *Arabidopsis*, disrupt seed dormancy and stress responses. *Hira-1* mutant seeds exhibited significantly reduced germination efficiency, especially under aged (controlled deterioration) or high salinity conditions, highlighting the importance of proper chromatin assembly for germination success.

Seed dormancy transition requires a switch from a repressive state to one permissive for germination, which is orchestrated by intricate epigenetic mechanisms. Authier et al. (2021) revealed that the RdDM-mediated DNA methylation fine-tunes germination based on the maternal environment during seed development. This suggests a transgenerational epigenetic memory via RdDM, transmitting environmental cues and influencing offspring traits. Moreover, DNA methylation also plays a complex role in heat stress resilience (Malabarba et al., 2021). While mild stress has minimal impact, severe heat triggers expression changes in ABA-mediated heat response genes, highlighting the dynamic interplay between DNA methylation and environmental conditions. Cui et al. (2022) revealed dynamic DNA methylation patterns in the invasive *Mikania micrantha* during germination. They observed increased CHH methylation early on, potentially linked to germination. Interestingly, later stages exhibited active DNA demethylation, suggesting a combined role for cold response and antioxidant regulation in seed germination potential. These findings imply that environmental cues trigger the de-repression of germination genes.

Wang et al. (2023) identified the histone H3K27 demethylase REF6 as a key positive regulator of light-induced seed germination in *Arabidopsis*. REF6 promotes the expression of light-responsive genes involved in diverse hormonal pathways and cellular processes, highlighting its role in coordinating germination events (Wang et al., 2023). Remarkably, REF6 influences germination through multiple hormonal pathways including gibberellin (GA), abscisic acid (ABA), and auxin signaling. These findings suggest a strong link between hormonal signaling and epigenetic control in modulating seed germination. Moreover, Zheng et al. (2022) shed light on how HDA15 fine-tunes seed germination by targeting GA biosynthesis. Through direct repression of GA-related genes, this histone deacetylase modulates the downstream effects of photoreceptor phytochrome B (PHYB) signaling, ultimately regulating the switch from dormancy to germination. Taken together, these results provide strong support that different histone modifications act in an orchestrated manner to shape the expression of germination-related genes through the cross-talk between environmental (e.g., light) and hormonal signaling pathways.

Chromatin remodelers regulate tridimensional genome architecture and nucleosome positioning, further enhancing/suppressing DNA accessibility for transcription (Bieluszewski et al., 2023). In *Arabidopsis*, DELAY OF GERMINATION1 (DOG1), a master regulator of primary dormancy (the dormancy stage in seeds dispersed from the mother plant), acts to delay germination by encoding a temperature detector that directs dormancy cycling upon seasonal variation (Ding et al., 2022). The DOG1 protein also acts as a key integrator of ABA signaling by physically interacting with PP2C phosphatases AHG1 and AHG3. This interaction suppresses ABA catabolism, resulting in enhanced ABA sensitivity and the imposition of primary seed dormancy (Carrillo-Barral et al., 2020). Polycomb group proteins deposit H3K27me3 marks on the *DOG1* promoter, silencing its expression (Chen et al., 2020). LUX, a circadian clock component, recruits the PICKLE (PKL) chromatin remodeling factor to the *DOG1* chromatin region and thus promote H3K27me3 deposition (Zhang et al., 2008; Zhang et al., 2012b; Ding et al., 2022). PKL directly associates with LUX, which binds to the regulatory elements of *DOG1*, ultimately mediating circadian regulation of *DOG1* expression (Zha et al., 2020). Non-coding RNAs can act as guides for these chromatin remodeling complexes, ensuring precise control over gene expression during germination. Interestingly, it has been shown that the core binding site of LUX is near the transcriptional start site of the non-coding antisense transcript *asDOG1*, known to suppress *DOG1* expression during seed maturation (Fedak et al., 2016).

Environmental cues also tightly regulate seed dormancy and germination through small non-coding RNAs. MicroRNAs play a critical role in this transition via RNA interference (RNAi), fine-tuning gene expression and influencing dormancy release (Sanan-Mishra & Kumari, 2020). Jiang et al. (2023) demonstrated the positive role of miR27319 in seed dormancy using *Arabidopsis* and rice lines manipulated for miR27319 expression. The authors revealed a high-temperature-mediated dormancy pathway regulated by miR27319 through its influence on phytohormone biosynthesis, ABA, and GA pathways. Jian et al. (2016) identified diverse miRNAs in *Brassica napus* seeds that regulate responses to salt or drought stress during imbibition, the initial stage of germination marked by water uptake. Putative targets for these miRNAs included stress-responsive genes like drought-responsive family proteins and stress-related proteins, suggesting a role for miRNA-mediated regulation in stress tolerance during germination (Jian et al., 2016).

Seed germination is intricately linked to hormonal signaling via ncRNA pathways. Liu et al. (2007) demonstrated that miR160 negatively regulates *AUXIN RESPONSE FACTOR10* (*ARF10*), impacting seed germination and development. This ncRNA (miR160) targets *ARF10*, influencing the interplay between auxin and ABA pathways. Transgenic plants expressing a mutant ARF10 resistant to miR160 (*mARF10*) displayed developmental defects, highlighting the critical role of miR160-ARF10 interaction in these processes. Moreover, Sarkar Das et al. (2018) revealed a regulatory network involving miRNAs and *trans*-acting siRNAs (tasiRNAs) in *Arabidopsis* seed germination. miR390 targets transcripts derived from the *TAS3 locus*, influencing the tasiR-ARF pathway that regulates *AUXIN RESPONSE FACTORS* (*ARF2/3/4*) expression. Interestingly, these findings suggest that the interplay between dynamic miRNA expression, their targets, and crosstalk with the ta-siRNA pathway contributes to germination control (Sarkar Das et al., 2018). Beyond their regulatory roles, small RNAs also contribute to seed dormancy and germination through selective mRNA oxidation and degradation (El-Maarouf-Bouteau et al., 2013). Unraveling the intricacies of non-coding RNA function in seed germination holds immense potential for elucidating the genetic basis of complex plant traits, thereby paving the way for advancements in crop productivity.

Transgenerational Plasticity and Seed Germination

Transgenerational epigenetic marks triggered by environmental factors can imprint heritable marks on genomes, impacting traits in offspring (Akhter et al., 2021). These epigenetic modifications may potentially ensure proper germination timing in response to seasonal changes, ultimately enhancing their overall fitness within specific ecological niches. For instance, drought in the parent generation can affect water stress responses in the germination of subsequent progeny seeds through heritable epigenetic marks. Consequently, drought-induced epigenetic changes in germline cells can be transmitted to offspring seeds, influencing their water stress responses and potentially priming them for future drought events (Nguyen et al., 2022). The inherited epigenetic states may potentially allow plants to fine-tune their responses to environmental challenges faced by their ancestors, in this case optimizing water usage under drought conditions.

Harnessing Dormancy and Germination in Crop Improvement

Epigenetic modifications dynamically evolve during development, marking crucial transitions like the switch from dormancy to active germination. Understanding the interplay between epigenetic programming and this transition will open additional ways to enhance crop resilience and productivity. However, deciphering the crosstalk between epigenetics and the seed dormancy-to-germination transition presents several challenges. Researchers aimed at exploring epigenetic mechanisms involved in such a switch should carefully select time points for sampling, a process that can inadvertently miss crucial phases due to limitations in resolution. Emerging high-throughput screening methods to profile the whole transcriptome of individual cells such as single-cell RNA-seq offer promising avenues for exploration, but their full adaptation for investigating cellular responses in seeds is still ongoing.

Environmental factors experienced by parent plants may potentially trigger transgenerational epigenetic effects on their seeds and offspring over a finite number of generations, ultimately impacting (and adding noise for understanding) dormancy and germination processes. Elucidating the presence of stable epialelles and differentiating them from modifications triggered by unstable, environmentally-driven epigenetic changes is challenging, often requiring controlled environments and multigenerational experiments. Furthermore, interpreting the functional phenotypic relevance of such epigenetic changes in the dormancy to germination transition is complex, as epigenetic marks exhibit context-dependent effects on gene expression. A deep understanding of seed biology is essential for deciphering these relationships.

Controlled growth environments, while they can mimic some aspects of field conditions, cannot replicate their full complexity. This introduces additional challenges for studying epigenetic factors, as the natural environment is tremendously complex and may significantly vary in its modes of action, differentially regulating seed dormancy and germination. Seeds during the dormancy-to-germination transition experience a spectrum of environmental challenges, further complicating the elucidation of their epigenetic responses. These challenges represent *per se* another layer of complexity to understanding how environmental factors influence epigenetic mechanisms. Even after identifying specific epigenetic changes, we should understand their functional significance. Distinguishing which modifications directly influence seed behavior and how they carry out these processes is complex and demands careful analysis.

Epigenetic modifications hold promise for enhancing desired seed traits and fostering crop resilience. Tailoring epigenetic marks associated with seed dormancy and germination could enable precise control over emergence timing, a crucial issue in crops under challenging climate change conditions. Seed priming, a pre-germination treatment using chemical, physical, or biological agents, induces controlled stress during early germination stages. This enhances synchronized emergence, seedling establishment, plant growth, and ultimately yield, particularly under stress conditions (Srivastava et al., 2021). Similarly, an "epigenetic" seed priming strategy could enhance seed stress tolerance during dormancy and germination, allowing plants to thrive under challenging environments. This approach would offer exciting possibilities for developing robust seeds that adapt effectively to a changing climate and compete successfully for resources.

Conclusions

Understanding and manipulating seed dormancy and germination becomes critical for ensuring crop success in the context of changing climate. By deciphering how epigenetic pathways influence genes controlling dormancy and germination, we can potentially unlock strategies for fine-tuning emergence timing. This could lead to crops emerging under optimal conditions, improving seedling establishment and overall yield. Furthermore, epigenetic modifications hold promise for enhancing seed stress tolerance during dormancy and germination. By priming seeds to activate these mechanisms, we can develop plants better adapted to face drought, salinity, or extreme temperatures—challenges that are likely to become more prevalent under climate change. Consequently, novel insights into epigenetics of seed dormancy and germination will shed light in generating climate-resilient crops, a critical step for ensuring food security upon challenging conditions.

References

Akhter, Z., Bi, Z., Ali, K., Sun, C., Fiaz, S., Haider, F. U., & Bai, J. (2021). In response to abiotic stress, DNA methylation confers epigenetic changes in plants. *Plants, 10*(6), 1096.

Authier, A., Cerdán, P., & Auge, G. A. (2021). Role of the RNA-directed DNA Methylation pathway in the regulation of maternal effects in Arabidopsis thaliana seed germination. *Micro Publication Biology*, 2021. https://doi.org/10.17912/micropub.biology.000504

Benhamed, M., Bertrand, C., Servet, C., & Zhou, D. X. (2006). Arabidopsis GCN5, HD1, and TAF1/HAF2 interact to regulate histone acetylation required for light-responsive gene expression. *The Plant Cell, 18*(11), 2893–2903.

Bieluszewski, T., Prakash, S., Roulé, T., & Wagner, D. (2023). The role and activity of SWI/SNF chromatin remodelers. *Annual Review of Plant Biology, 74*, 139–163.

Carrillo-Barral, N., Rodríguez-Gacio, M. D. C., & Matilla, A. J. (2020). Delay of Germination-1 (DOG1): A key to understanding seed dormancy. *Plants, 9*(4), 480.

Chen, N., Wang, H., Abdelmageed, H., Veerappan, V., Tadege, M., & Allen, R. D. (2020). HSI2/VAL1 and HSL1/VAL2 function redundantly to repress DOG1 expression in Arabidopsis seeds and seedlings. *The New Phytologist, 227*, 840–856. https://doi.org/10.1111/nph.16559

Cui, C., Wang, Z., Su, Y., & Wang, T. (2022). Antioxidant regulation and DNA methylation dynamics during Mikania micrantha seed germination under cold stress. *Frontiers in Plant Science, 13*, 856527.

Ding, X., Jia, X., Xiang, Y., & Jiang, W. (2022). Histone modification and chromatin remodeling during the seed life cycle. *Frontiers in Plant Science, 13*, 865361.

El-Maarouf-Bouteau, H., Meimoun, P., Job, C., Job, D., & Bailly, C. (2013). Role of protein and mRNA oxidation in seed dormancy and germination. *Frontiers in Plant Science, 4*, 77.

Fedak, H., Palusinska, M., Krzyczmonik, K., Brzezniak, L., Yatusevich, R., Pietras, Z., et al. (2016). Control of seed dormancy in Arabidopsis by a cis-acting noncoding antisense transcript. *Proceedings of the National Academy of Sciences, 113*(48), E7846–E7855.

Gomez-Cabellos, S., Toorop, P. E., Cañal, M. J., Iannetta, P. P., Fernández-Pascual, E., Pritchard, H. W., & Visscher, A. M. (2022). Global DNA methylation and cellular 5-methylcytosine and H4 acetylated patterns in primary and secondary dormant seeds of Capsella bursa-pastoris (L.) Medik.(shepherd's purse). *Protoplasma, 259*, 595–614.

Gu, D., Ji, R., He, C., Peng, T., Zhang, M., Duan, J., et al. (2019). Arabidopsis histone methyltransferase SUVH5 is a positive regulator of light-mediated seed germination. *Frontiers in Plant Science, 10*, 456773.

Han, Y., Georgii, E., Priego-Cubero, S., Wurm, C. J., Hüther, P., Huber, G., et al. (2023). Arabidopsis histone deacetylase HD2A and HD2B regulate seed dormancy by repressing DELAY OF GERMINATION 1. *Frontiers in Plant Science, 14*, 1124899.

Jian, H., Wang, J., Wang, T., Wei, L., Li, J., & Liu, L. (2016). Identification of rapeseed microRNAs involved in early stage seed germination under salt and drought stresses. *Frontiers in Plant Science, 7*, 658.

Jiang, H., Gao, W., Jiang, B. L., Liu, X., Jiang, Y. T., Zhang, L. T., et al. (2023). Identification and validation of coding and non-coding RNAs involved in high-temperature-mediated seed dormancy in common wheat. *Frontiers in Plant Science, 14*, 1107277.

Layat, E., Bourcy, M., Cotterell, S., Zdzieszyńska, J., Desset, S., Duc, C., et al. (2021). The Histone chaperone HIRA is a positive regulator of seed germination. *International Journal of Molecular Sciences, 22*(8), 4031.

Liu, P. P., Montgomery, T. A., Fahlgren, N., Kasschau, K. D., Nonogaki, H., & Carrington, J. C. (2007). Repression of AUXIN RESPONSE FACTOR10 by microRNA160 is critical for seed germination and post-germination stages. *The Plant Journal, 52*(1), 133–146.

Luján-Soto, E., & Dinkova, T. D. (2021). Time to wake up: Epigenetic and small-RNA-mediated regulation during seed germination. *Plants, 10*(2), 236.

Malabarba, J., Windels, D., Xu, W., & Verdier, J. (2021). Regulation of DNA (de) methylation positively impacts seed germination during seed development under heat stress. *Genes, 12*(3), 457.

Narváez, G., Muñoz-Espinoza, C., Soto, E., Rothkegel, K., Bastías, M., Gutiérrez, J., et al. (2022). Global methylation analysis using MSAP reveals differences in chilling-associated DNA methylation changes during dormancy release in contrasting sweet cherry varieties. *Horticulturae, 8*(10), 962.

Nguyen, N. H., Vu, N. T., & Cheong, J. J. (2022). Transcriptional stress memory and transgenerational inheritance of drought tolerance in plants. *International Journal of Molecular Sciences, 23*(21), 12918.

Rothkegel, K., Sánchez, E., Montes, C., Greve, M., Tapia, S., Bravo, S., et al. (2017). DNA methylation and small interference RNAs participate in the regulation of MADS-box genes involved in dormancy in sweet cherry (Prunus avium L.). *Tree Physiology, 37*(12), 1739–1751.

Sanan-Mishra, N., & Kumari, A. (2020). Role of RNA interference in seed germination. *Plant Small RNA*, 101–116.

Sarkar Das, S., Yadav, S., Singh, A., Gautam, V., Sarkar, A. K., Nandi, A. K., et al. (2018). Expression dynamics of miRNAs and their targets in seed germination conditions reveals miRNA-ta-siRNA crosstalk as regulator of seed germination. *Scientific Reports, 8*(1), 1233.

Srivastava, A. K., Suresh Kumar, J., & Suprasanna, P. (2021). Seed 'primeomics': Plants memorize their germination under stress. *Biological Reviews, 96*(5), 1723–1743.

Wang, Y., Gu, D., Deng, L., He, C., Zheng, F., & Liu, X. (2023). The histone H3K27 demethylase REF6 is a positive regulator of light-initiated seed germination in Arabidopsis. *Cells, 12*(2), 295.

Zha, P., Liu, S., Li, Y., Ma, T., Yang, L., Jing, Y., & Lin, R. (2020). The evening complex and the chromatin-remodeling factor PICKLE coordinately control seed dormancy by directly repressing DOG1 in Arabidopsis. *Plant Communications, 1*(2), 100011.

Zhang, H., Rider, S. D., Jr., Henderson, J. T., Fountain, M., Chuang, K., Kandachar, V., et al. (2008). The CHD3 remodeler PICKLE promotes trimethylation of histone H3 lysine 27. *The Journal of Biological Chemistry, 28*, 22637–22648. https://doi.org/10.1074/jbc.M802129200

Zhang, L., Wang, Y., Zhang, X., Zhang, M., Han, D., Qiu, C., & Han, Z. (2012a). Dynamics of phytohormone and DNA methylation patterns changes during dormancy induction in strawberry (Fragaria× ananassa Duch.). *Plant Cell Reports, 31*, 155–165.

Zhang, H., Bishop, B., Ringenberg, W., Muir, W. M., & Ogas, J. (2012b). The CHD3 remodeler PICKLE associates with genes enriched for trimethylation of histone H3 lysine 27. *Plant Physiology, 159*, 418–432. https://doi.org/10.1104/pp.112.194878

Zhao, M., Yang, S., Liu, X., & Wu, K. (2015). Arabidopsis histone demethylases LDL1 and LDL2 control primary seed dormancy by regulating DELAY OF GERMINATION 1 and ABA signaling-related genes. *Frontiers in Plant Science, 6*, 159.

Zheng, F., Wang, Y., Gu, D., & Liu, X. (2022). Histone deacetylase HDA15 restrains PHYB-dependent seed germination via directly repressing GA20ox1/2 gene expression. *Cells, 11*(23), 3788.

Zhou, C., Zhang, L., Duan, J., Miki, B., & Wu, K. (2005). HISTONE DEACETYLASE19 is involved in jasmonic acid and ethylene signaling of pathogen response in Arabidopsis. *Plant Cell, 17*, 1196–1204.

Zhou, Y., Tan, B., Luo, M., Li, Y., Liu, C., Chen, C., Yu, C. W., Yang, S., Dong, S., Ruan, J., et al. (2013). HISTONE DEACETYLASE19 interacts with HSL1 and participates in the repression of seed maturation genes in Arabidopsis seedlings. *Plant Cell, 25*, 134–148.

Zhou, Y., Yang, P., Zhang, F., Luo, X., & Xie, J. (2020). Histone deacetylase HDA19 interacts with histone methyltransferase SUVH5 to regulate seed dormancy in Arabidopsis. *Plant Biology, 22*(6), 1062–1071.

Part IV

Epigenetics: A Fast and Versatile Response to Climate Change

Chapter 9
Epigenetic Mechanisms and Stress Tolerance in a Climate Change Scenario

Environmental stresses, including drought, extreme temperatures, soil salinity, and nutrient deficiencies, pose significant challenges to global agriculture. The intensification of these stressors due to climate change exacerbates food security concerns worldwide. To address these challenges, understanding the molecular mechanisms underlying plant adaptation is crucial. Epigenetic mechanisms offer a promising avenue for developing resilient crops, as they provide a flexible and responsive layer of gene regulation.

Epigenetics and Environmental Response

The role of epigenetic mechanisms in shaping plant responses to environmental stress is a cornerstone in our understanding of ecological resilience. By orchestrating rapid and reversible changes in gene expression, epigenetic mechanisms allow plants to adapt in dynamic and often hostile environments. Consequently, our knowledge of the underlying epigenetic mechanisms acting upon environmental stress can be the key to our food systems' future. Crops exhibit specific epigenetic variants that have already been shown to contribute to their tolerance to abiotic stresses (Samantara et al., 2021). By identifying sources of epigenetic variation, we can significantly increase crop stress tolerance, safeguarding food production in unpredictable environments.

L. M. Vaschetto, *Epigenetics in Crop Improvement*,
https://doi.org/10.1007/978-3-031-73176-1_9

DNA Methylation

During stress responses, plants can experience changes in global DNA methylation levels. In *Arabidopsis*, Dowen et al. (2012) demonstrated that dynamic DNA methylation patterns within repetitive sequences, including transposable elements (TEs), act as a molecular switch, controlling the expression of neighboring genes in response to biotic stress, thus offering a perspective on stress responses. Van Antro et al. (2023) further observed transgenerational inheritance of temperature-induced DNA methylation marks in clonal duckweed (*Lemna minor* L.) lineages, enabling gene expression profiles to persist across generations. Remarkably, this "epigenetic memory"is enriched in particular sequence contexts such as those observed in TEs (e.g., repeat arrays), suggesting a sequence-biased transgenerational mechanism for stress adaptation acting at the epigenetic level. Bioinformatic tools integrating biological omics data is crucial for deciphering how to exploit nucleotide sequence contexts targeted by specific epigenetic enzymes/complexes in molecular breeding programs. The nature of plant DNA methylation patterns across plant species under stress is highly dynamic, where the direction and magnitude of change depend on the specific stressor and its duration over time (Pazzaglia et al., 2023).

Increasing evidence supports the idea that dynamic DNA methylation changes enable plants to prioritize stress adaptation over growth in challenging environments. For instance, global DNA hypomethylation leads to upregulation of stress-responsive genes, including those involved in osmotic regulation. Wada et al. (2004) linked stress-responsive gene upregulation in tobacco (*Nicotiana tabacum L.*) to hypomethylation, suggesting that dynamic DNA methylation plays a role in stress response. In plants, active DNA demethylation is mediated by members of the enzyme family ROS1/DME (Liu & Lang, 2020). Through recognition and removal of methylated cytosines, stress-responsive signaling pathways initiate a demethylation cascade, replacing epigenetic modifications and altering gene expression patterns. For instance, it has been shown that the abscisic acid (ABA) signaling pathway activates DNA demethylases that promote the demethylation of stress-responsive genes (Kim et al., 2019). As a result, these genes become more accessible for transcription, potentially facilitating rapid and enhanced stress responses upon subsequent stress exposure. The passive demethylation mechanism occurs during cellular replication when DNA is not targeted by DNA methyltransferases (Saze et al., 2003). Moreover, stress-induced demethylation marks may potentially persist transgenerationally, thereby priming offspring for enhanced stress tolerance even in the absence of the initial stressor. Both passive and active DNA demethylation mechanisms may be crucial for stress responses. Conversely, DNA hypermethylation of genes related to growth and development can potentially repress them, conserving energy resources during stress. Sun et al. (2021) demonstrated that DNA methylation plays a role in desiccation tolerance and stress memory in the model plant *Boea hygrometrica*. They observed acclimation-induced DNA hypermethylation in the CHH context, coinciding with changes in genes regulating methylation. Moreover, DNA hypermethylation in the CHH context is associated with a high expression of stress-responsive genes under salinity stress in the salt-tolerant rice variety Pokkali (Rajkumar et al., 2020). These findings suggest that DNA methylation plays a key role in plant adaptation to dehydration, with the interplay between DNA methylation and gene expression offering a potential mechanism for plants to adapt to challenging climate conditions.

Stress-induced DNA methylation changes are not random but intricately linked to canonical (genetically defined) pathways. Upon stress perception by cellular sensors (e.g., drought or salt receptors), a signaling cascade culminates in targeted gene expression changes induced by epigenetic modifications. For instance, it has been shown that the stress hormone abscisic acid (ABA) induces DNA methylation changes in grapevine, promoting phenolic accumulation and enhancing red wine quality (Marfil et al., 2019). The integration of DNA methylation profiles into cellular signaling pathways allows for precise and coordinated regulation of both stress-responsive and non-stress genes. These stress-induced DNA methylation profiles may potentially be heritable through meiosis, enabling transgenerational epigenetic inheritance. The progeny of stressed plants can thus exhibit enhanced stress tolerance even without the initial stressor, suggesting transmission of "stress memory"across generations and a potential survival advantage. This type of stress memory may be mediated by distinct epigenetic mechanisms, including DNA methylation and histone modifications (Gallusci et al., 2023).

Stress-induced DNA demethylation may potentially exert multiple effects in enhanced stress tolerance. By demethylating stress-responsive genes, DNA demethylation may become more accessible the transcriptional machinery, facilitating gene expression responses to stressors. This phenomenon could allow plants to allocate resources effectively, prioritizing stress adaptation, and optimizing nutrient utilization under challenging conditions. Furthermore, this mechanism may also contribute to stress memory, enabling plants to "remember" prior stress exposure. The memory effect would result in faster and more efficient responses upon subsequent stress encounters, which may be particularly advantageous in a climate change scenario, where stress events are sporadic and unpredictable. This transgenerational aspect underscores the significance of epigenetic modifications in plant adaptation to challenging environmental conditions.

Histone Modifications

Histone modifications have shown to alter chromatin structure to contribute to the transcriptional regulation of stress-responsive genes, enabling plants to face environmental challenges. In *Arabidopsis*, Khan et al. (2020) observed that the protein Powerdress (PWR) binds to the Histone Deacetylase 9 (HDA9) upon dehydration stress, leading to altered acetylation of the *CYP707A1* gene, a key player in the ABA catabolic pathway. The PWR-HDA9 complex regulates chromatin modifications, altering the expression of genes involved in ABA signaling and ultimately shaping the plant's response to ABA and drought stress. Plants frequently exhibit dynamic changes in histone acetylation/deacetylation in response to abiotic stresses. These modifications may exert direct control over the expression of stress-responsive genes. Generally, histone acetylation within the promoter regions is associated with increased gene expression. Zhao et al. (2014) showed that promoter-associated histone acetylation at H3K9 and H4K5 residues is involved in the stress-induced transcriptional regulation of the osmotic stress responsive maize gene *dehydration-responsive element binding protein 2A* (*ZmDREB2A*). Roy et al. (2014) further observed that differential acetylation of multiple histone H3 marks (H3K9, H3K14, and H3K27) within the regulatory region of the rice *DREB1* ortholog, *OsDREB1b*, promotes chromatin remodeling and transcriptional activation during cold stress. Similarly, Yolcu et al. (2016) linked salt stress to transcriptional activation of a *peroxidase* (*POX*) gene in *Beta vulgaris* and *Beta maritima*. This activation coincided with increased acetylation at H3K9 and H3K27 histone marks, suggesting a role for these histone modifications in stress-responsive gene regulation. Consequently, acetylation marks likely allow easier access of the transcriptional machinery to DNA, promoting gene expression. Conversely, under stress conditions, histone deacetylation can lead to reduced transcription (Liu et al., 2014). The selective suppression of genes associated with growth and development represents *per se* an adaptive response, prioritizing stress adaptation over other biological processes.

Histone methylation also plays a pivotal role in influencing chromatin structure and gene expression under stress conditions. Different methylation patterns at target histone residues can have distinct effects on gene expression. Generally, trimethylation of H3K4 (H3K4me3) is associated with gene transcription (Chen et al., 2015), while H3K27me3 is linked to gene repression (Derkacheva et al., 2013; Tan et al., 2022), although there are exceptions to these trends (Foroozani et al., 2021). Similarly to acetylation, histone methylation has been shown to mediate stress adaptation by influencing gene expression (Bobadilla & Berr, 2016). Consequently, the interplay between histone-modifying and DNA methylation enzymes allows plants to fine-tune their transcriptional responses to abiotic stressors, optimizing survival and adaptation under different environmental conditions.

Non-coding RNAs

Epigenetic regulation extends beyond DNA methylation and histone modifications, with non-coding RNAs (ncRNAs) adding another layer of complexity to the control of stress-responsive genes, which is particularly noteworthy under challenging conditions such as those imposed by climate change. Optimizing the regulatory potential of ncRNAs offers a promising breeding strategy for developing crops better equipped to tolerate environmental fluctuations. By understanding how ncRNAs fine-tune gene expression patterns during stress responses, we can identify key regulatory elements. These genetic sequences can then be harnessed through targeted breeding or (epi)genetic engineering. For instance, breeders can select plants

exhibiting naturally occurring variations in ncRNAs that enhance their stress tolerance. Alternatively, genetic engineering techniques could be used to precisely
silence ncRNAs and thus modulate target specific genes. This targeted approach can
potentially lead to the development of crops with enhanced resilience to drought,
salinity, extreme temperatures, and other environmental stressors associated with
climate change.

lncRNAs

Long non-coding RNAs (lncRNAs) are key regulators in diverse developmental
processes and stress responses across plants. Functioning at transcriptional, post-
transcriptional, and epigenetic levels, lncRNAs influence stress-responsive genes
directly and indirectly by targeting regulatory genes encoding transcription factors
and ncRNAs (e.g., miRNAs) that control gene networks (Jha et al., 2020). Stress-
induced lncRNAs have been identified in diverse crop species including, among
others, foxtail millet (*Setaria italica*), sorghum (*Sorghum bicolor*), maize (*Zea
mays*), rice (*Oryza sativa*), switchgrass (*Panicum virgatum*), and wheat (*Triticum
aestivum*). In foxtail millet and sorghum, intergenic lncRNAs (lincRNAs) and natural antisense lncRNAs (NAT-lncRNAs) exhibiting differential gene expression patterns under drought stress have been identified (Qi et al., 2013). Chen et al. (2021)
revealed a novel drought stress response pathway in rice mediated by a lncRNA
(TCONS_00021861). This lncRNA potentially act as a competing endogenous
RNA (ceRNA), sponging miR528-3p and thereby enhancing target mRNA expression, ultimately leading to auxin overproduction and stress tolerance (Chen et al.,
2021). Zhang et al. (2018) identified differentially expressed lncRNAs in switchgrass (*Panicum virgatum L.*) upon repeated dehydration stress. These lncRNAs targeted genes involved in key stress response pathways, including hormonal (ABA
and ethylene) biosynthesis, signal transduction, and carbohydrate metabolism.
Remarkably, the study also revealed lncRNAs potentially linked to stress (dehydration) memory, which are associated with chromatin epigenetic modifications, particularly histone methylation (Zhang et al., 2018). In hexaploid wheat, Ren et al.
(2018) identified differentially expressed circular RNAs (circRNAs) associated
with low nitrogen (LN) stress. Some of the identified circRNAs were predicted to
function as miRNA sponges, potentially capturing miRNAs and preventing them
from silencing their target genes. This suggests a role for circRNAs in regulating
root system development during LN stress, as enhanced root growth allows plants
to explore a larger soil volume for nitrogen acquisition (Ren et al., 2018).

Small Non-coding RNAs

Small *non-coding* RNAs (sncRNAs) enhance abiotic stress tolerance by silencing Transposable Elements (TEs) involved in stress signaling (Wang & Chekanova, 2016). Papareddy et al. (2020) investigated the biogenesis and function of small interfering RNAs (siRNAs) derived from TEs in *Arabidopsis* embryos. They observed that decondensation of heterochromatin in response to environmental factors appears to promote siRNA production and function. The study also revealed that these siRNAs are essential for directing the re-establishment of DNA methylation on the originating TEs, thereby maintaining their silencing (Papareddy et al., 2020). In barley, Hackenberg et al. (2015) revealed that drought stress selectively regulates the expression of miRNAs and other small RNAs (sRNAs), including repeat-associated small interfering RNAs (rasiRNAs) derived from genomic repetitive elements like retrotransposons and heterochromatic regions. Remarkably, low-expressed miRNAs and rasiRNAs tended to be downregulated. Interestingly, the identified ncRNA targets were primarily associated with transcriptional regulation (Hackenberg et al., 2015). NcRNAs exert additional control over stress responses by forming regulatory modules through interactions with various proteins. Gu et al. (2023) elucidated a critical siRNA-dependent regulatory module involved in rice thermotolerance. SUPPRESSOR OF GENE SILENCING 3a (OsSGS3a) is a protein that interacts with its homolog OsSGS3b to modulate the biogenesis of *trans*-acting siRNAs (tasiRNAs) targeting AUXIN RESPONSE FACTORS (ARFs). RDR6 and Dicer-like (DCL) proteins process double-stranded RNAs (dsRNAs) into mature tasiRNAs, regulating different transcription factors, including MYBs, HEAT-INDUCED TAS1 TARGETS (HTTs), and ARFs. Remarkably, they observed that the OsSGS3a-dependent tasiRNA-OsARF3 module acts as a positive regulator of thermotolerance, ultimately controlling the trade-off between abiotic and biotic stress responses (Gu et al., 2023).

MiRNAs act as master regulators that mediate plant tolerance to a broad spectrum of abiotic stresses linked to climate change, including drought, temperature extremes, nutrient deficiency, UV radiation, oxidative stress, and hypoxia. Stress-induced microRNAs (miRNAs) selectively silence specific stress-responsive genes that could potentially have deleterious consequences (Ding et al., 2020). These gene-silencing mechanisms act as a buffer against excessive stress damage. Kuczyński et al. (2022) identified over 150 differentially expressed miRNAs in four soybean cultivars upon exposure to cold stress. The study found that soybean miRNAs potentially regulate plant abiotic stress responses by targeting genes involved in reactive oxygen species scavenging, flavonoid biosynthesis, and osmotic potential regulation (Kuczyński et al., 2022). In sweetpotato (*Ipomoea batatas*), Yu et al. (2020) observed that both chilling and heat stress alter the expression of stress-responsive miRNAs, suggesting their potential role in regulating plant response in seedlings. Additionally, both stresses induced the activity of enzymes involved in oxidative stress, indicating the activation of a cellular defense mechanism against oxidative damage (Yu et al., 2020). Zhang et al. (2019) revealed that heat stress significantly alters miRNA expression in maize plants, with over 60 identified miR-NAs being downregulated. They identified antagonistic miRNA-mRNA interactions, suggesting miRNA silencing of these genes during stress response. Remarkably, targeted mRNAs included transcription factors, protein kinases, and signaling components, highlighting the potential for broad-scale modulation of cellular processes under heat stress (Zhang et al., 2019).

Developing Resilient Crops

Resilience is a biological term frequently utilized in the context of climate change. It refers to the capacity of a given species to withstand disturbances, adapt to changing conditions, and maintain essential functionality and structural features despite external pressures. Under ever-changing environments, resilience becomes a key factor for measuring the adaptability of crops. Climate change introduces several abiotic stressors, including, among others, altered temperatures and precipitation patterns, and increased occurrence of extreme events (e.g., heatwaves, tropical cyclones, etc.). These stressors can seriously affect food security and human livelihoods. Therefore, enhancing plant resilience is a critical matter to ensure the continued functioning and stability of crop systems. Epigenetic mechanisms foster resilience in crops by regulating the expression of stress-responsive genes, enhancing their environmental tolerance. Studies have shown that stressed plants experience resilience-associated epigenetic modifications, some of which can then be inherited by their offspring. For instance, in rice, Kou et al. (2011) provided evidence that acquired adaptive traits induced by nitrogen stress may be potentially heritable. This phenomenon has been linked to inheritable changes in DNA methylation, suggesting an epigenetic basis for transgenerational inheritance. By promoting the transmission of adaptive epigenetic marks, plant populations can increase their resilience and maintain ecological functions in altered landscapes.

Adaptive phenotypic plasticity is an evolutionary mechanism that allows plants to respond to challenging environmental conditions within their lifetime. This flexibility allows plants to optimize their fitness, adjusting their morphology, physiology, and behavior in response to stressful conditions. It is also a biological process that provides a rapid means of adaptation, allowing individual plants to thrive in their current environment. By facilitating the survival and reproduction of individuals under changing conditions, adaptive phenotypic plasticity can contribute to the persistence of crops in the context of climate change. Some environmentally driven evolutionary strategies associated with adaptive phenotypic plasticity include:

- Temperature fluctuations: Plants exhibit plastic responses to temperature fluctuations, including altered flowering times, changes in photosynthetic rates, etc.
- Drought Stress: Plastic responses to drought are diverse and include reduced leaf area, increased root-to-shoot ratios, and altered stomata integrity. These adapta-

tions help plants conserve water and maintain their physiological functions during water scarcity.

- Altered Precipitation Patterns: Plants exhibit adaptive responses, such as adjustments in stomatal opening and closure, to face unpredictable precipitation events.

Epigenetics, Phenotypic Plasticity, and Adaptative Strategies

Epigenetic mechanisms provide a means by which plants fine-tune plastic adaptive responses to specific environmental conditions. Drought, salinity, and extreme temperatures can trigger epigenetic modifications associated with altered gene expression, enabling plants to adapt their physiology and morphology to withstand such conditions (Kim et al., 2015). Additionally, epigenetic marks can potentially be transmitted as a memory of past environmental exposures, influencing the adaptive responses of subsequent generations and expediting adaptation to recurring stressors (Mirbahai & Chipman, 2014). Future breeding strategies may involve pre-exposing seeds to controlled stress conditions to establish stress memory in plants, ultimately enhancing their resilience through priming applications.

To understand how plants face the challenges presented by a changing climate, it is essential to explore the dynamic interplay between epigenetic inheritance, adaptive phenotypic plasticity, and long-term adaptation strategies. Adaptive phenotypic plasticity is a short-term strategy employed by plants to respond promptly to immediate environmental changes. When faced with altered conditions, like increased temperatures or reduced water availability, plants can adjust their physiological features to optimize their chances of survival and reproduction (McDowell et al., 2008). For instance, in response to prolonged drought, plants can reduce their leaf size, increase root elongation, or alter their flowering times. This plasticity allows them to face unpredictable and fluctuating environmental stressors. These changes may potentially be orchestrated by epigenetic modifications that modulate gene expression to match the new environmental conditions. Phenotypic plasticity can then allow plants to rapidly adapt to changing circumstances, increasing their chances of survival and reproduction. While adaptive phenotypic plasticity provides a rapid response to environmental challenges, transgenerational epigenetic inheritance could play a pivotal role in long-term adaptation. In such a case, epigenetic modifications should then be passed on to subsequent generations, providing a memory that primes offspring for similar challenges.

The interplay between transgenerational epigenetic inheritance and adaptive phenotypic plasticity would underlie epigenetic marks acquired through plastic responses, inherited by subsequent generations, that ultimately would lead to an adaptive advantage (Jablonka, 2013). This integration of short-term plasticity and long-term epigenetic inheritance would allow for a robust adaptive strategy. In a changing climate scenario, rapid adaptation is critical for long-term survival. Hence, epigenetics and phenotypic variation could provide an appropriate framework for rapid adaptation. This crosstalk might allow plants to adjust gene expression in favor of desired phenotypic traits, enabling them to thrive in previously hostile environmental conditions.

Limitations and Perspectives

Climate change is characterized by increased environmental variability, including changes in temperature, precipitation, and atmospheric gases (CO_2/O_2) levels. This variability poses difficulties in isolating and studying specific environmental factors and their effects on epigenetic mechanisms. Epigenetic modifications are also highly variable, and usually non-heritable. The immense majority of epigenetic changes are triggered by environmental changes in a cell-specific manner, confined to individual cell lineages, and not passed on to offspring. Discriminating between transient, cell-specific epigenetic marks and those associated with environmentally driven transgenerational heritability is a challenging task. Moreover, establishing causality between these epigenetic modifications and phenotypic changes also remains challenging. Correlation between epigenetic modifications and environmentally modifiable traits does not necessarily imply causation. The epigenetic landscape is extremely complex, involving many different types of chemical modifications and their synergistic interactions. Deciphering this complexity and understanding which epigenetic marks are functionally relevant in response to climate change is a formidable challenge for molecular breeders.

Comparative epigenetics involves studying epigenetic variation across different individuals in response to varying environmental conditions. This approach can be useful for studying adaptive-plastic phenotypes in the context of climate change (Navarro-Martín et al., 2020). By comparing the epigenetic landscape of plants that are more or less resilient to environmental variation, we can identify specific epigenetic marks associated with crop resilience. These marks can then be used as targets for breeding or engineering more resilient varieties. It requires developing epigenetically tractable models and manipulation of specific chemical marks to assess their causal roles in phenotypic plasticity and crop adaptation, which may aid in establishing causality. Integrating epigenetics with various omics approaches such as transcriptomics, proteomics, and metabolomics can further provide a holistic view of how epigenetic changes translate into functional phenotypic adaptations in crops.

Conclusion

Understanding the intricate interplay between epigenetic mechanisms, phenotypic plasticity, and adaptive strategies is a major issue for developing resilient crops capable of thriving in the face of climate change. By harnessing epigenetic modifications, ncRNAs and enzymes with potential to modify the epigenetic landscape, we can unlock the potential to rapidly adapt crops to fluctuating environmental conditions. Integrating these insights into breeding programs is instrumental in ensuring global food security and mitigating the impacts of climate change on agriculture.

References

Bobadilla, R., & Berr, A. (2016). Histone methylation-a cornerstone for plant responses to environmental stresses. In A. Shanker & C. Shanker (Eds.), *Abiotic and biotic stress in plants-recent advances and future perspectives* (pp. 31–61). IntechOpen.

Chen, X., Liu, X., Zhao, Y., & Zhou, D. X. (2015). Histone H3K4me3 and H3K27me3 regulatory genes control stable transmission of an epimutation in rice. *Scientific Reports, 5*(1), 13251.

Chen, J., Zhong, Y., & Qi, X. (2021). LncRNA TCONS_00021861 is functionally associated with drought tolerance in rice (Oryza sativa L.) via competing endogenous RNA regulation. *BMC Plant Biology, 21*, 1–12.

Derkacheva, M., Steinbach, Y., Wildhaber, T., Mozgova, I., Mahrez, W., Nanni, P., et al. (2013). Arabidopsis MSI1 connects LHP1 to PRC2 complexes. *The EMBO Journal, 32*(14), 2073–2085.

Ding, Y., Huang, L., Jiang, Q., & Zhu, C. (2020). MicroRNAs as important regulators of heat stress responses in plants. *Journal of Agricultural and Food Chemistry, 68*(41), 11320–11326.

Dowen, R. H., Pelizzola, M., Schmitz, R. J., Lister, R., Dowen, J. M., Nery, J. R., et al. (2012). Widespread dynamic DNA methylation in response to biotic stress. *Proceedings of the National Academy of Sciences, 109*(32), E2183–E2191.

Foroozani, M., Vandal, M. P., & Smith, A. P. (2021). H3K4 trimethylation dynamics impact diverse developmental and environmental responses in plants. *Planta, 253*, 4. https://doi.org/10.1007/s00425-020-03520-0

Gallusci, P., Agius, D. R., Moschou, P. N., Dobránszki, J., Kaiserli, E., & Martinelli, F. (2023). Deep inside the epigenetic memories of stressed plants. *Trends in Plant Science, 28*(2), 142–153.

Gu, X., Si, F., Feng, Z., et al. (2023). The OsSGS3-tasiRNA-OsARF3 module orchestrates abiotic-biotic stress response trade-off in rice. *Nature Communications, 14*, 4441. https://doi.org/10.1038/s41467-023-40176-2

Hackenberg, M., Gustafson, P., Langridge, P., & Shi, B. J. (2015). Differential expression of micro RNA s and other small RNA s in barley between water and drought conditions. *Plant Biotechnology Journal, 13*(1), 2–13.

Jablonka, E. (2013). Epigenetic inheritance and plasticity: The responsive germline. *Progress in Biophysics and Molecular Biology, 111*(2–3), 99–107.

Jha, U. C., Nayyar, H., Jha, R., et al. (2020). Long non-coding RNAs: Emerging players regulating plant abiotic stress response and adaptation. *BMC Plant Biology, 20*, 466. https://doi.org/10.1186/s12870-020-02595-x

Khan, I. U., Ali, A., Khan, H. A., Baek, D., Park, J., Lim, C. J., et al. (2020). PWR/HDA9/ABI4 complex epigenetically regulates ABA dependent drought stress tolerance in Arabidopsis. *Frontiers in Plant Science, 11*, 623.

Kim, J. M., Sasaki, T., Ueda, M., Sako, K., & Seki, M. (2015). Chromatin changes in response to drought, salinity, heat, and cold stresses in plants. *Frontiers in Plant Science, 6*, 114.

Kim, J. S., Lim, J. Y., Shin, H., Kim, B. G., Yoo, S. D., Kim, W. T., & Huh, J. H. (2019). ROS1-dependent DNA demethylation is required for ABA-inducible NIC3 expression. *Plant Physiology, 179*(4), 1810–1821.

Kou, H. P., Li, Y., Song, X. X., Ou, X. F., Xing, S. C., Ma, J., et al. (2011). Heritable alteration in DNA methylation induced by nitrogen-deficiency stress accompanies enhanced tolerance by progenies to the stress in rice (Oryza sativa L.). *Journal of Plant Physiology, 168*(14), 1685–1693.

Kuczyński, J., Gracz-Bernaciak, J., Twardowski, T., Karłowski, W. M., & Tyczewska, A. (2022). Cold stress-induced miRNA and degradome changes in four soybean varieties differing in chilling resistance. *Journal of Agronomy and Crop Science, 208*(6), 777–794.

Liu, R., & Lang, Z. (2020). The mechanism and function of active DNA demethylation in plants. *Journal of Integrative Plant Biology, 62*(1), 148–159.

Liu, X., Yang, S., Zhao, M., Luo, M., Yu, C. W., Chen, C. Y., et al. (2014). Transcriptional repression by histone deacetylases in plants. *Molecular Plant, 7*(5), 764–772.

Marfil, C., Ibañez, V., Alonso, R., Varela, A., Bottini, R., Masuelli, R., et al. (2019). Changes in grapevine DNA methylation and polyphenols content induced by solar ultraviolet-B radiation, water deficit and abscisic acid spray treatments. *Plant Physiology and Biochemistry, 135*, 287–294.

McDowell, N., Pockman, W. T., Allen, C. D., Breshears, D. D., Cobb, N., Kolb, T., et al. (2008). Mechanisms of plant survival and mortality during drought: Why do some plants survive while others succumb to drought? *New Phytologist, 178*(4), 719–739.

Mirbahai, L., & Chipman, J. K. (2014). Epigenetic memory of environmental organisms: A reflection of lifetime stressor exposures. *Mutation Research/Genetic Toxicology and Environmental Mutagenesis, 764*, 10–17.

Navarro-Martín, L., Martyniuk, C. J., & Mennigen, J. A. (2020). Comparative epigenetics in animal physiology: An emerging frontier. *Comparative Biochemistry and Physiology Part D: Genomics and Proteomics, 36*, 100745.

Papareddy, R. K., Páldi, K., Paulraj, S., et al. (2020). Chromatin regulates expression of small RNAs to help maintain transposon methylome homeostasis in Arabidopsis. *Genome Biology, 21*, 251. https://doi.org/10.1186/s13059-020-02163-4

Pazzaglia, J., Dattolo, E., Ruocco, M., Santillán-Sarmiento, A., Marin-Guirao, L., & Procaccini, G. (2023). DNA methylation dynamics in a coastal foundation seagrass species under abiotic stressors. *Proceedings of the Royal Society B, 290*(1991), 20222197.

Qi, X., Xie, S., Liu, Y., Yi, F., & Yu, J. (2013). Genome-wide annotation of genes and noncoding RNAs of foxtail millet in response to simulated drought stress by deep sequencing. *Plant Molecular Biology, 83*, 459–473.

Rajkumar, M. S., Shankar, R., Garg, R., & Jain, M. (2020). Bisulphite sequencing reveals dynamic DNA methylation under desiccation and salinity stresses in rice cultivars. *Genomics, 112*(5), 3537–3548.

Ren, Y., Yue, H., Li, L., Xu, Y., Wang, Z., Xin, Z., & Lin, T. (2018). Identification and characterization of circRNAs involved in the regulation of low nitrogen-promoted root growth in hexaploid wheat. *Biological Research, 51*, 43.

Roy, D., Paul, A., Roy, A., Ghosh, R., Ganguly, P., & Chaudhuri, S. (2014). Differential acetylation of histone H3 at the regulatory region of OsDREB1b promoter facilitates chromatin remodelling and transcription activation during cold stress. *PLoS One, 9*(6), e100343.

Samantara, K., Shiv, A., de Sousa, L. L., Sandhu, K. S., Priyadarshini, P., & Mohapatra, S. R. (2021). A comprehensive review on epigenetic mechanisms and application of epigenetic modifications for crop improvement. *Environmental and Experimental Botany, 188*, 104479.

Saze, H., Scheid, O. M., & Paszkowski, J. (2003). Maintenance of CpG methylation is essential for epigenetic inheritance during plant gametogenesis. *Nature Genetics, 34*(1), 65–69.

Sun, R. Z., Liu, J., Wang, Y. Y., & Deng, X. (2021). DNA methylation-mediated modulation of rapid desiccation tolerance acquisition and dehydration stress memory in the resurrection plant Boea hygrometrica. *PLoS Genetics, 17*(4), e1009549.

Tan, F. Q., Wang, W., Li, J., Lu, Y., Zhu, B., Hu, F., et al. (2022). A coiled-coil protein associates Polycomb repressive complex 2 with KNOX/BELL transcription factors to maintain silencing of cell differentiation-promoting genes in the shoot apex. *The Plant Cell, 34*(8), 2969–2988.

Van Antro, M., Prelovsek, S., Ivanovic, S., Gawehns, F., Wagemaker, N. C., Mysara, M., et al. (2023). DNA methylation in clonal duckweed (Lemna minor L.) lineages reflects current and historical environmental exposures. *Molecular Ecology, 32*(2), 428–443.

Wada, Y., Miyamoto, K., Kusano, T., & Sano, H. (2004). Association between up-regulation of stress-responsive genes and hypomethylation of genomic DNA in tobacco plants. *Molecular Genetics and Genomics, 271*, 658–666.

Wang, H. L. V., & Chekanova, J. A. (2016). Small RNAs: Essential regulators of gene expression and defenses against environmental stresses in plants. *Wiley Interdisciplinary Reviews: RNA, 7*(3), 356–381.

Yolcu, S., Ozdemir, F., Güler, A., & Bor, M. (2016). Histone acetylation influences the transcriptional activation of POX in Beta vulgaris L. and Beta maritima L. under salt stress. *Plant Physiology and Biochemistry, 100*, 37–46.

Yu, J., Su, D., Yang, D., Dong, T., Tang, Z., Li, H., et al. (2020). Chilling and heat stress-induced physiological changes and microRNA-related mechanism in sweetpotato (Ipomoea batatas L.). *Frontiers in Plant Science, 11*, 687.

Zhang, C., Tang, G., Peng, X., et al. (2018). Long non-coding RNAs of switchgrass (Panicum virgatum L.) in multiple dehydration stresses. *BMC Plant Biology, 18*, 79. https://doi.org/10.1186/s12870-018-1288-3

Zhang, M., An, P., Li, H., Wang, X., Zhou, J., Dong, P., et al. (2019). The miRNA-mediated post-transcriptional regulation of maize in response to high temperature. *International Journal of Molecular Sciences, 20*(7), 1754.

Zhao, L., Wang, P., Yan, S., Gao, F., Li, H., Hou, H., et al. (2014). Promoter-associated histone acetylation is involved in the osmotic stress-induced transcriptional regulation of the maize ZmDREB2A gene. *Physiologia Plantarum, 151*(4), 459–467.

Chapter 10
Epimutations and Metastable Epialleles: Exploring a Mine of Hidden Variation in Crops

For many years, the focus of plant breeders and researchers striving to enhance crop yield, resilience, and quality has revolved around genetic diversity at the DNA sequence level. This approach, rooted in traditional genetics, has yielded invaluable insights and improvements. However, in recent years, we have discovered hidden additional layers of epigenetic diversity, underscoring the importance of understanding variation from a holistic view. Epigenetic diversity comes in the form of epimutations and metastable epialleles that offer a new dimension to explore. These epigenetic modifications can exert a significant effect on gene expression and the resulting phenotypes, making them targetable for crop improvement. Even more important is the fact that epigenetic marks are generally not static, which means they can be reversibly influenced by environmental factors, stressors, and developmental cues. Epialleles can be meiotically heritable, passing from one generation to the next through the germline, and even revert to their original state over a few generations. This dynamic nature becomes vital for crops to thrive under fluctuating environmental conditions.

Epialleles and Epimutations

The identification and manipulation of epigenetic variants hold an immense potential for improving plant traits associated with adaptation and stress responses. Epialleles can be defined as epigenetic modifications, mainly arisen during early epigenetic programming, linked to differential gene expression patterns in genetically identical individuals (Dolinoy et al., 2007). Naturally occurred epialleles have

L. M. Vaschetto, *Epigenetics in Crop Improvement*,
https://doi.org/10.1007/978-3-031-73176-1_10

been reported to affect critical adaptive traits such as flowering time (Cubas et al., 1999), flower development (Luo et al., 1996), fruit ripening (Manning et al., 2006), and even sex determination (Martin et al., 2009). Table 10.1 summarizes stable epialleles already identified and characterized. These epialleles have arisen from either spontaneous natural processes or exposure to mutagenic agents.

The term "epimutations" refers to stable/persistent modifications in the epigenetic landscape of specific genes or genomic regions that alter gene expression but do not involve changes in the underlying DNA sequence (Cini et al., 2015). These

Table 10.1 Epialleles identified in crops (modified from Gahlaut et al., 2020)

Species	Gene/*locus*	Affected Trait	Epitrigger	Reference
Arabidopsis thaliana	*SUP* (*SUPERMAN*)	Flower morphology	Mutagenic agent	Jacobsen and Meyerowitz (1997)
	AG (*AGAMOUS*)	Flower morphology	Mutagenic agent	Jacobsen et al. (2000)
	FWA (*Flowering Wageningen*)	Flowering time	Mutagenic agent	Soppe et al., 2000
	FOLT1 (*folate transporter 1*)	Fertility	*Trans*-acting ncRNA	Durand et al. (2012)
	PPH (*Pheophytin Pheophorbide Hydrolase*)	Leaf senescence	Spontaneous mutagenesis	He et al. (2018)
	HISN6B (*Histidinol-phosphate aminotransferase 1*)	Hybrid incompatibility	Spontaneous mutagenesis	Blevins et al. (2017)
	HEI10 (*HUMAN ENHANCER OF CELL INVASION NO.10*)	Plant adaptation to climate change	Spontaneous mutagenesis	Chen et al. (2024)
	BNS (*BONSAI*)	Growth	Mutation in the *ddm1* gene	Saze et al. (2008)
	BAL1	Disease resistance	Mutagenic agent	Stokes et al. (2002)
Zea Mays	*R1* (*red*)	Pigmentation	Spontaneous mutagenesis	Brink (1956)
	B1 (*booster 1*)	Pigmentation	Spontaneous mutagenesis	Patterson et al. (1993)
	PL1 (*purple plant*)	Pigmentation	Spontaneous mutagenesis	Hollick et al. (1995)
	P1 (*pericarp color*)	Pigmentation	Spontaneous mutagenesis	Cocciolone et al. (2001)
Lcyc (Linaria cycliodea)	*Lcyc*	Flower morphology	Spontaneous mutagenesis	Cubas et al. (1999)
Solanum lycopersicum	*CNR* (*Colorless non-ripening*)	Ripening	Spontaneous mutagenesis	Manning et al. (2006)
	VTE3 (*Vitamin E*)	Vitamin E content	Spontaneous mutagenesis	Quadrana et al. (2014)

(continued)

Table 10.1 (continued)

Species	Gene/*locus*	Affected Trait	Epitrigger	Reference
Oryza sativa	*D1 (Drawf1)*	Height	Spontaneous mutagenesis	Miura et al. (2009)
	FIE1 (Fertilization-Independent Endosperm 1)	Height	Spontaneous mutagenesis	Zhang et al. (2012)
	RAV6 [Related to Abscisic Acid Insensitive 3 (ABI3)/Viviparous1 (VP1) 6]	Leaf morphology and seed	Spontaneous mutagenesis	Zhang et al. (2015)
	AK1 (Adenylate Kinase)	Photosynthesis	Spontaneous mutagenesis	Wei et al. (2017)
	ESP (Epigenetic Short Panicle)	Panicle morphology	Spontaneous mutagenesis	Luan et al. (2019)
	SPL14 (Squamosa Promoter binding protein-Like)	Panicle morphology and seed	Spontaneous mutagenesis	Miura et al. (2010)
Brassica rapa	*SP11/SCR (S locus protein 11/S locus cystein rich)*	Self-incompatibility	*Trans*-acting ncRNA	Shiba et al. (2006)
Elaeis guineensis	*DEF1 (DEFICIENS)*	Mantled fruit	Spontaneous mutagenesis (transposon methylation)	Ong-Abdullah et al. (2015)
Cucumis melo	*CmWIP1 (WASP/N--WASP-interacting protein 1)*	Sex determination	Transposon insertion	Martin et al. (2009)

epigenetic modifications can manifest either spontaneously or in response to environmental stimuli. In plants, epimutations have been reported to produce metastable epialleles capable of affecting important phenotypic traits such as, for example, stress tolerance. For instance, *Arabidopsis* plants exposed to high-salinity stress may exhibit a dramatic increase of DNA methylation at CG sites, with approximately 75% of these modifications being transmitted to subsequent generations, although some may be progressively lost (Jiang et al., 2014). Zheng et al. (2017) demonstrated that multi-generational drought exposure triggers transgenerational epimutations, enabling rice plants to adapt to subsequent drought conditions. To fully decipher the potential of epigenetic diversity, it is crucial to understand the mechanisms that govern the generation of epimutations and their stability over time. In other words, elucidating the existence and impact of epimutations offers vital insights for enhancing crop traits and exploiting their diversity.

Metastable epialleles can be defined as epigenetically modifiable genes/*loci* in a variable and reversible manner (Jirtle & Skinner, 2007). The concept of metastable epialleles highlights the interplay between genetics and epigenetics in shaping individual variation. These alleles may be established early in life and exhibit epigenetic-driven variability in expression, making them particularly susceptible to environmental influences and potentially contributing to complex phenotypes (Dolinoy et al., 2007). The existence of metastable epialleles challenges the traditional idea of genetic determinism, as clonal (genetically identical) individuals can exhibit distinct phenotypes due to variation in their epigenetic marks.

Unveiling the molecular basis of epimutations is one of the hottest topics in epigenetics. Consistent with observations in *Arabidopsis* for epialleles in *SUPERMAN* (*SUP*) and *AGAMOUS* (*AG*) (Jacobsen & Meyerowitz, 1997; Jacobsen et al., 2000), as well as the spontaneous epi-allele of the *Lcyc* gene in *Linaria vulgaris* (Cubas et al., 1999), epimutations often exhibit extensive methylation across the gene bodies, leading to its silencing in the mutant state (Soppe et al., 2000). In contrast to this pattern, *fwa* exhibits gain-of-function epimutations characterized by a complete lack of cytosine methylation within two directly repeated sequences positioned in contraposition in both the promoter and coding regions of the gene. This absence of methylation contrasts with DNA methylation observed in the wild-type *FWA* allele (Soppe et al., 2000).

Epigenetic Variation in Cereal Crops

Maize is a model system for studying epialleles and epigenetic phenomena, including imprinting, paramutation, and transgenerational epigenetic inheritance. These phenomena have revealed how plants establish and maintain heritable variation patterns beyond DNA sequence and contrasting to Mendel's conclusions. A prime example of this is paramutation at the *b1 locus*, which encodes a key anthocyanin biosynthesis activator (Haring et al., 2010; Goettel & Messing, 2013). The functional *B-I* and silenced *B'* epialleles of the *b1 locus* have identical DNA sequences but differ in their methylation patterns within a 7-repeat upstream regulatory region (853 bp each). Paramutation at the maize *pericarp color 1* (*p1*) gene, encoding a Myb-like transcription factor responsible for phlobaphene biosynthesis, also serves to exemplify the inherent complexity of stable epialleles. While extensive methylation analysis revealed no differences beyond a 17 kb upstream region (Goettel & Messing, 2013), the key distinction between expressed and silenced alleles resides within a complex Transposable Element (TE)-rich sequence. This suggests that haplotype-specific epigenetic variation regarding the nucleotide sequence context (i.e., TE insertion), might underlie the existence of epigenetically regulated alleles. Sidorenko et al. (2024) shed light on the role of transcribed enhancers in paramutation of the maize *p1* gene. The *P1-rr* allele undergoes paramutation upon exposure to a transgene carrying its enhancer fragment (1.2-kb). Fine-mapping revealed that the nucleotide sequence essential for paramutation reside within this fragment, overlapping a previously identified enhancer present in four direct repeats of *P1-rr*. Remarkably, the paramutated *P1-rr'* allele (light-pigmented) exhibits higher small RNA levels compared to the active *P1-rr* state. These findings suggest that *P1-rr* enhancer repeats play a crucial role in mediating paramutation, potentially through small RNA-based silencing pathways.

Rice (*Oryza sativa*) contains several identified epialleles, being also an useful resource to explore the underlying phenomenon of transgenerational epigenetic inheritance. Zhang et al. (2012) identified a dominant gain-of-function epiallele (*Epi-df*) of the *FERTILIZATION-INDEPENDENT ENDOSPERM1* (*FIE1*) gene, linked to developmental defects. *FIE1*'s dual role as a Polycomb repressive complex 2 (PRC2) component for H3K27me3-mediated silencing and maternally expressed gene in the endosperm makes it a prime epigenetic target. Remarkably, *Epi-df* exhibits hypomethylation, reduced H3K9me2, and increased H3K4me3 levels (Zhang et al., 2012). Epigenetic reprogramming disrupted *FIE1* imprinting and altered H3K27me3 levels in hundreds of genes, highlighting the link between epiallele-mediated epigenetic changes and developmental phenotypes. Zhang et al. (2015) identified a functionally important epiallele, *Epi-rav6*, in rice. This gain-of-function variant of the *RELATED TO ABI3/VP1 6* (*RAV6*) gene led to abnormal *RAV6* expression associated with hypomethylation in its promoter region. Interestingly, a conserved miniature inverted-repeat transposable element (MITE) insertion within this region plays a key role in regulating *RAV6* expression. *Epi-rav6* plants display altered expression of genes involved in brassinosteroid (BR) signaling and biosynthesis, resulting in reduced grain size (Zhang et al., 2015). This study highlights the potential of TE-mediated epialleles to influence agronomic traits and suggests their widespread presence in crop genomes. Wei et al. (2017) revealed an epiallele, *Epi-ak1*, that silences the *OsAK1* gene, encoding a key adenosine phosphate enzyme. Hypermethylation within the *OsAK1* promoter triggers this silencing, leading to downregulation of photosynthesis genes and a decrease in photosynthetic rates. Consequently, *Epi-ak1* plants exhibit reduced grain number and yield. Beyond DNA methylation, ncRNAs have also been implicated in epiallele variation in rice. Luan et al. (2019) identified *Epi-sp*, a gain-of-function epiallele of the rice *ESP locus* encoding a putative lncRNA. DNA hypomethylation in the *ESP* promoter led to its abnormal expression, causing the dense and short panicle phenotype observed in *Epi-sp* plants (Luan et al., 2019).

Epialleles, Evolution, and Adaptation

Epialleles have shown to play critical roles during plant evolution and adaptation. Durand et al. (2012) shed light on the potential role of epialleles in incompatibility mechanisms, highlighting epialleles as potential drivers of genetic divergence following gene duplication events. The authors of this study observed that a TE-mediated rearrangement in *Arabidopsis thaliana* led to a duplicated *FOLT1* gene. One copy acquired DNA methylation and silenced the other, demonstrating epiallele stability across generations (Durand et al., 2012). Consequently, they proposed that the rearranged gene triggers RNA-directed DNA methylation, silencing the native copy and promoting genetic incompatibility after gene duplication (Durand et al., 2012). Similarly, Martin et al. (2009) revealed a TE-induced epiallele in the *CmWIP1* gene, governing sex determination in melon. The inserted transposon triggers DNA methylation across the *CmWIP1* promoter, leading to its expression and carpel abortion. *CmWIP1* also indirectly represses *CmACS-7*, facilitating stamen development and ultimately, unisexual male flower formation. These findings reveal that TE insertions play a critical role in generating epialleles, potentially contributing to the diversification leading to speciation events.

Epialleles and Climate Change

Epialleles have already shown their value for breeding programs aiming to develop climate-resilient crops. In *Arabidopsis*, Chen et al. (2024) identified an epiallele (*HEI10*), characterized by differential DNA methylation and linked to climate-responsive traits. This epiallele highlights the association between epiallelic variation, carbon dioxide (CO_2) emissions -a key climate change parameter-, and potential plant adaptation through natural selection (Chen et al., 2024). Similarly, Saban et al. (2020) explored the transgenerational effects of elevated CO_2 on *Plantago lanceolata L.* They observed minimal genetic differentiation but significant divergence in DNA methylation profiles between populations exposed to high CO_2 levels for multiple generations and a control site, highlighting the potential for transgenerational plasticity of DNA methylation profiles in response to long-term CO_2 exposure.

Epigenetic Recombinant Inbred Lines (epiRILs)

Epigenetic Recombinant Inbred Lines (epiRILs) can be defined as a collection of plants that differ for their epigenetic profiles (e.g., DNA methylation) and show no genetic variation. EpiRILs are valuable tools for dissecting the role of epigenetics in environmentally responsive traits (e.g., stress tolerance). *Decreased DNA Methylation I (DDM1)* is a plant gene encoding for a nucleosome-remodeling ATPase that maintains methylation. Johannes et al. (2009) generated *Arabidopsis* epiRILs by crossing a *ddm1* mutant with wild-type plants (Fig. 10.1), revealing heritable variation in flowering time, plant height, and DNA methylation patterns across generations. These epiRILs displayed variation and high heritability for flowering time and plant height, while stably inheriting parental epialleles for at least eight generations (Johannes et al., 2009). Also in *Arabidopsis*, Reinders et al. (2009) employed a similar approach to generate epiRILs, crossing a mutant for methyltransferases (*met1-3*) and thereby lacking DNA methylation with its wild type, resulting in randomized epiallele distribution. They investigated inheritance patterns and phenotypic consequences of this epigenetic variation. Remarkably, TEs, previously inactive, exhibited mobility in over a quarter of the epiRILs. This finding suggests that strategically inactivating enzymes with epigenetic functions hold promise as a strategy to activate TEs capable of introducing potentially

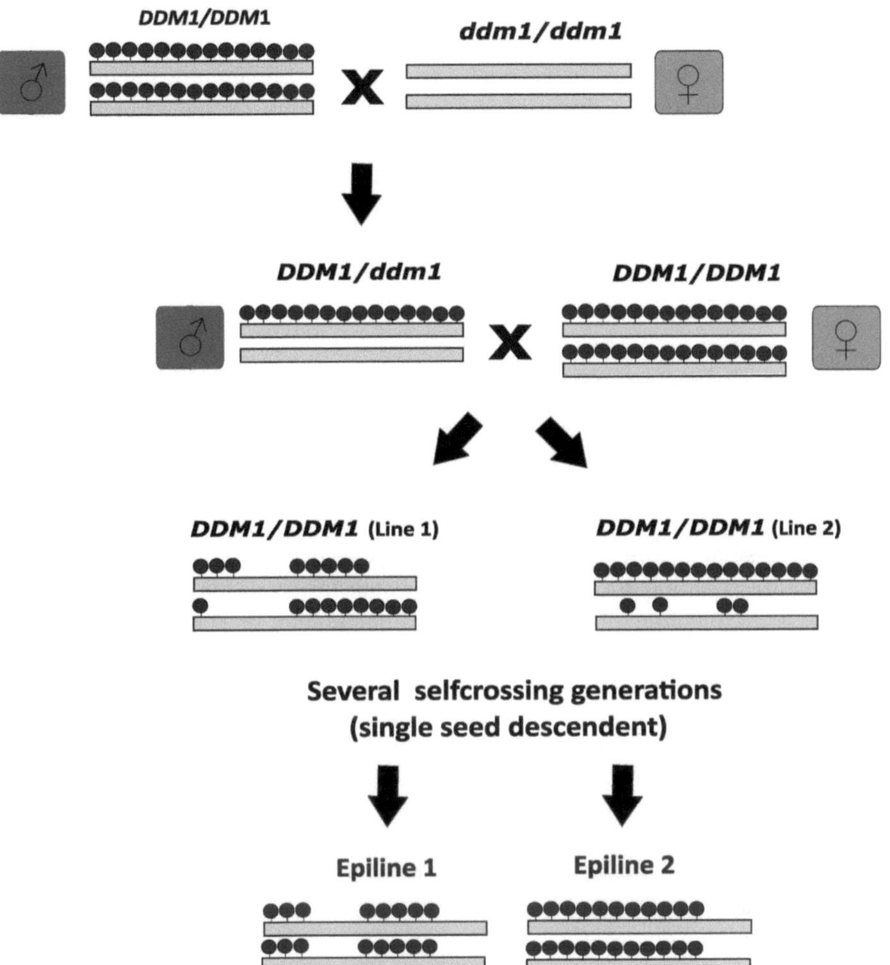

Fig. 10.1 Construction of plant epiRIL populations. (Adapted from Johannes et al., 2009). Yellow bars represent the diploid genome of the plant. Blue circles represent normal DNA methylation levels across the genome. The parental plants are isogenic plants except for the *Deficient in DNA Methylation 1* (*DDM1*) gene, which encodes an evolutionary conserved nucleosome remodeler that facilitates DNA methylation and silencing repetitive elements. In this model, the male parental line has the wild-type (*DDM1*) allele and exhibits normal methylation levels, while the female parent is a *ddm1* mutant (female) with globally reduced methylation. A hybrid F1 individual is backcrossed to the wt *DDM* parental line, and then the resulting lines obtained from this cross are selfed per several generations using a single seed descent methodology. This renders in the obtention of genetically identical epiRILs (Line 1, Line 2, etc.) with distinct methylation levels at specific *loci*. These "epilines" allow researchers to study how methylation stably affects plant traits across generations

heritable variations, offering a valuable tool for molecular breeding programs. Additionally, some traits, including flowering time and stress responses, displayed variation potentially linked to the *met1-3* parental line, highlighting the influence of epialleles on phenotypic diversity and genome stability (Reinders et al., 2009). Interestingly, both *Arabidopsis* epiRIL populations displayed contrasting phenotypic stability. Reinders et al. (2009) observed highly unstable phenotypes in *met1-3* lines, likely due to the lack of DNA methylation. In contrast, Johannes et al. (2009) found a significant stability (>99% of lines) in *ddm1*-derived epiRILs even after eight generations, suggesting efficient maintenance of epialleles.

Inducing Epigenetic Variation

Both environmental stressors and genetic mutations can induce epimutations that hold potential for crop improvement. As mentioned above, mutations in genes encoding epigenetic regulators, such as chromatin remodelers, DNA methylation enzymes, histone modifiers, or small RNA biogenesis machinery, can disrupt epiallele stability. Additionally, dysfunction in this machinery can occur spontaneously, leading to epimutations. Understanding the functioning of epimutation-mediated triggers is essential for exploiting their potential when developing epigenetically driven strategies for improving crop traits. By elucidating the drivers of epimutations, researchers can develop strategies to induce, stabilize, and inherit beneficial epigenetic modifications. Consequently, this knowledge is essential for exploiting the potential of epimutations in enhancing complex crop traits, including yield, stress tolerance, or nutritional quality.

Metastable epialleles are characterized by inter-individual variability in epigenetic states within a population. These epigenetically mediated alleles share a unique characteristic: their epigenetic marks are not universally stable across all individuals but rather exhibit a degree of variability. This diversity can result in distinct phenotypes among genetically identical individuals. Therefore, metastable epialleles can be distinguished by their differential heritability across generations, with the epigenetic states being passed on to offspring at variable rates. The stability of metastable epialleles and induced epimutations is dependent on different factors such as the presence of specific epigenetic marks on the genome region where the epigenetic alterations are localized, parental-specific crossing, etc. Environmental factors may further modulate the stability of epigenetic modifications. Temperature, for example, can upregulate the expression of DNA demethylases, potentially leading to the removal of epigenetic marks (Dutta et al., 2022). The intricate interplay between molecular and environmental factors governs the elusive stability of epialleles, posing a challenge to unlocking their dynamic potential in gene regulation mechanisms. Future advancements will likely pave the way for deciphering this crosstalk, allowing targeted manipulation of epigenetic states for agricultural applications.

Crop Epigenetics: Nurturing the Future of Agriculture

As our planet experiences rapid changes in climate patterns, crop production faces unprecedented challenges. Rising temperatures, irregular precipitation cycles, and unpredictable extreme weather events threaten food security, thereby we urgently need to develop adaptable and resilient crops capable of withstanding such adverse conditions. Although crops have been bred and adapted to specific climatic conditions over centuries, now face environmental conditions that may be beyond their historical range of variability. Epigenetics represents an unexplored field to improve agricultural traits in contemporaneous ever-changing environments, with a central role in terms of shaping the future of sustainable agriculture. Introgressing "epivariability" present in natural wild-relative crop populations, especially those present in species growing in extreme conditions, will be a cornerstone when developing crops with better performance.

Epigenetic variation, with its multi-level capacity to influence gene expression and phenotypic plasticity, offers an invaluable tool for improving traits in crops under challenging environments. By modifying the epigenetic landscape, we can potentially adapt crop growth to specific conditions and enhance their resilience to climate-induced stressors. Scientists have already designed tools to target genes responsible for development and stress tolerance and modify their epigenetic states (Vaschetto, 2018). Alternatively, it is also feasible potentiate crop productivity by epigenetically modulating crop genes associated with pathogen defense responses or silencing them in pathogen genomes (Vaschetto, 2021).

Despite the benefits of harnessing epigenetic variation for reversible trait modification, several challenges remain to be addressed. One critical limitation involves achieving a precise and predictable control of epigenetic changes under desired conditions. Ensuring that the epigenetic modifications occur without unintended consequences requires targetable sequence-specific systems and a deep understanding of the underlying epigenetic regulatory mechanisms. Confirming the heritability of the epigenetic changes is also of paramount importance, as only stably inherited epigenetic changes ensure trait modification. Molecular breeders should be aware to control epigenetic modifications, ultimately avoiding unintended consequences. Epigenomic engineering tools capable of effectively reprograming the epigenetic landscape are still in early stages, thereby improving their precision is a prerequisite for regular use in crop improvement.

The integration of epigenetics into molecular breeding programs also raises ethical questions. How should we consider the potential risks and benefits of epigenetically modified crops? Should epigenetic modifications be considered at the same regulatory level as genetic modifications? What are the environmental and health implications of manipulating epigenetic marks? These questions beyond the scope of this book require careful consideration and appropriate compliance mechanisms to ensure the responsible and sustainable use of epigenetic variability in molecular breeding.

To conclude, epigenetics holds immense potential for assisting in the development of next-generation crops. It offers a mechanism for rapidly enhancing crop resilience and adaptability, vital in a world where the climate is being increasingly unpredictable. However, it also demands a comprehensive approach, considering the intricacies of epigenetic regulation, the need for long-term stability, and regulatory dimensions. As we elucidate underlying molecular mechanisms and develop new tools to exploit epigenetics in crop improvement, we must ensure that manipulation of this type of variation is made responsibly and sustainably, safeguarding food security for the next generations.

References

Blevins, T., Wang, J., Pflieger, D., Pontvianne, F., & Pikaard, C. S. (2017). Hybrid incompatibility caused by an epiallele. *Proceedings of the National Academy of Sciences of the United States of America, 114*(14), 3702–3707.

Brink, R. A. (1956). A genetic change associated with the R locus in maize which is directed and potentially reversible. *Genetics, 41*, 872–889.

Chen, B., Wang, M., Guo, Y., Zhang, Z., Zhou, W., Cao, L., et al. (2024). Climate-related naturally occurring epimutation and their roles in plant adaptation in A. thaliana. *Molecular Ecology, e17356*.

Cini, G., Carnevali, I., Quaia, M., Chiaravalli, A. M., Sala, P., Giacomini, E., et al. (2015). Concomitant mutation and epimutation of the MLH1 gene in a lynch syndrome family. *Carcinogenesis, 36*(4), 452–458.

Cocciolone, S. M., Chopra, S., Flint-Garcia, S. A., McMullen, M. D., & Peterson, T. (2001). Tissue-specific patterns of a maize Myb transcription factor are epigenetically regulated. *The Plant Journal, 27*, 467–478.

Cubas, P., Vincent, C., & Coen, E. (1999). An epigenetic mutation responsible for natural variation in floral symmetry. *Nature, 401*, 157–161.

Dolinoy, D. C., Das, R., Weidman, J. R., & Jirtle, R. L. (2007). Metastable epialleles, imprinting, and the fetal origins of adult diseases. *Pediatric Research, 61*(7), 30–37.

Durand, S.; Bouche, N.; Elsa Perez, Strand, ; Loudet O.; Camilleri, C. Rapid establishment of genetic incompatibility through natural epigenetic variation. Current Biology 2012, 22, 326–331.

Dutta, M., Raturi, V., Gahlaut, V., Kumar, A., Sharma, P., Verma, V., et al. (2022). The interplay of DNA methyltransferases and demethylases with tuberization genes in potato (Solanum tuberosum L.) genotypes under high temperature. *Frontiers in Plant Science, 13*, 933740.

Gahlaut, V., Zinta, G., Jaiswal, V., & Kumar, S. (2020). Quantitative epigenetics: A new avenue for crop improvement. *Epigenomes, 4*(4), 25.

Goettel, W., & Messing, J. (2013). Epiallele biogenesis in maize. *Gene, 516*(1), 8–23.

Haring, M., Bader, R., Louwers, M., Schwabe, A., van Driel, R., & Stam, M. (2010). The role of DNA methylation, nucleosome occupancy and histone modifications in paramutation. *The Plant Journal, 63*(3), 366–378.

He, L., Wu, W., Zinta, G., Yang, L., Wang, D., Liu, R., Zhang, H., Zheng, Z., Huang, H., Zhang, Q., & Zhu, J. K. (2018). A naturally occurring epiallele associates with leaf senescence and local climate adaptation in Arabidopsis accessions. *Nature Communications, 9*(1), 460.

Hollick, J. B., Patterson, G. I., Coe, E. H., Cone, K. C., & Chandler, V. L. (1995). Allelic interactions heritably influence the activity of a metastable maize pl allele. *Genetics, 141*, 709–719.

Jacobsen, S. E., & Meyerowitz, E. M. (1997). Hypermethylated superman epigenetic alleles in Arabidopsis. *Science, 277*, 1100–1103.

Jacobsen, S. E., Sakai, H., Finnegan, E. J., Cao, X., & Meyerowitz, E. M. (2000). Ectopic hypermethylation of flower specific genes in Arabidopsis. *Current Biology, 24*, 179–186.

Jiang, C., Mithani, A., Belfield, E. J., Mott, R., Hurst, L. D., & Harberd, N. P. (2014). Environmentally responsive genome-wide accumulation of de novo Arabidopsis thaliana mutations and epimutations. *Genome Research, 24*(11), 1821–1829.

Jirtle, R. L., & Skinner, M. K. (2007). Environmental epigenomics and disease susceptibility. *Nature Reviews Genetics, 8*(4), 253–262.

Johannes, F., Porcher, E., Teixeira, F. K., Saliba-Colombani, V., Simon, M., Agier, N., et al. (2009). Assessing the impact of transgenerational epigenetic variation on complex traits. *PLoS Genetics, 5*(6), e1000530.

Luan, X., Liu, S., Ke, S., Dai, H., Xie, X. M., Hsieh, T. F., & Zhang, X. Q. (2019). Epigenetic modification of ESP, encoding a putative long noncoding RNA, affects panicle architecture in rice. *Rice, 12*(1), 2.

Luo, D., Carpenter, R., Vincent, C., Copsey, L., & Coen, E. (1996). Origin of floral asymmetry in antirrhinum. *Nature, 383*(6603), 794–799.

Manning, K., Tör, M., Poole, M., Hong, Y., Thompson, A. J., King, G. J., et al. (2006). A naturally occurring epigenetic mutation in a gene encoding an SBP-box transcription factor inhibits tomato fruit ripening. *Nature Genetics, 38*(8), 948–952.

Martin, A., Troadec, C., Boualem, A., Rajab, M., Fernandez, R., Morin, H., Pitrat, M., Dogimont, C., & Bendahmane, A. (2009). A transposon-induced epigenetic change leads to sex determination in melon. *Nature, 461*(7267), 1135–1138.

Miura, K., Agetsuma, M., Kitano, H., Yoshimura, A., Matsuoka, M., Jacobsen, S. E., & Ashikari, M. (2009). A metastable dwarf1 epigenetic mutant affecting plant stature in rice. *Proceedings of the National Academy of Sciences of the United States of America, 106*, 11218–11223.

Miura, K., Ikeda, M., Matsubara, A., Song, X. J., Ito, M., Asano, K., Matsuoka, M., Kitano, H., & Ashikari, M. (2010). Osspl14 promotes panicle branching and higher grain productivity in rice. *Nature Genetics, 42*, 545–549.

Ong-Abdullah, M., Ordway, J. M., Jiang, N., Ooi, S. E., Kok, S. Y., Sarpan, N., Azimi, N., Hashim, A. T., Ishak, Z., Rosli, S. K., et al. (2015). Loss of karma transposon methylation underlies the mantled somaclonal variant of oil palm. *Nature, 525*(7570), 533–537.

Patterson, G. I., Thorpe, C. J., & Chandler, V. L. (1993). 1993 Paramutation, an allelic interaction, is associated with a stable and heritable reduction of transcription of the maize b regulatory gene. *Genetics, 135*, 881–894.

Quadrana, L., Almeida, J., Asís, R., Duffy, T., Dominguez, P. G., Bermúdez, L., Conti, G., Corrêa da Silva, J. V., Peralta, I. E., Colot, V., Asurmendi, S., Fernie, A. R., Rossi, M., & Carrari, F. (2014). Natural occurring epialleles determine vitamin E accumulation in tomato fruits. *Nature Communications, 5*, 3027.

Reinders, J., Wulff, B. B., Mirouze, M., Marí-Ordóñez, A., Dapp, M., Rozhon, W., et al. (2009). Compromised stability of DNA methylation and transposon immobilization in mosaic Arabidopsis epigenomes. *Genes & Development, 23*(8), 939–950.

Saban, J. M., Watson-Lazowski, A., Chapman, M. A., & Taylor, G. (2020). The methylome is altered for plants in a high CO2 world: Insights into the response of a wild plant population to multigenerational exposure to elevated atmospheric [CO_2]. *Global Change Biology, 26*(11), 6474–6492.

Saze, H., Shiraishi, A., Miura, A., & Kakutani, T. (2008). Control of genic DNA methylation by a jmjc domain- containing protein in Arabidopsis thaliana. *Science, 319*, 462–465.

Shiba, H., Kakizaki, T., Iwano, M., Tarutani, Y., Watanabe, M., Isogai, A., & Takayama, S. (2006). Dominance relationships between self-incompatibility alleles controlled by DNA methylation. *Nature Genetics, 38*(3), 297–299.

Sidorenko, L. V., Chandler, V. L., Wang, X., & Peterson, T. (2024). Transcribed enhancer sequences are required for maize p1 paramutation. *Genetics, 226*(1), iyad178.

Soppe, W. J. J., Jacobsen, S. E., Alonso-Blanco, C., Jackson, J. P., Kakutani, T., Koornneef, M., & Peeters, A. J. M. (2000). The late flowering phenotype of fwa mutants is caused by gain-of-function epigenetic alleles of a homeodomain gene. *Molecular Cell, 2000*(6), 791–802.

Stokes, T. L., Kunkel, B. N., & Richards, E. J. (2002). Epigenetic variation in Arabidopsis disease resistance. *Genes & Development, 16*, 171–182.

Vaschetto, L. M. (2018). Modulating signaling networks by CRISPR/Cas9-mediated transposable element insertion. *Current Genetics, 64*(2), 405–412.

Vaschetto, L. M. (2021). *RNAi strategies for pest management: Methods and protocols.* Volume N° 2360. Hardcover ISBN: 978-1-0716-1632-1. Humana Press, Springer Science Business Media (Springer Nature). https://doi.org/10.1007/978-1-0716-1633-8

Wei, X., Song, X., Wei, L., Tang, S., Sun, J., Hu, P., & Cao, X. (2017). An epiallele of rice AK1 affects photosynthetic capacity. *Journal of Integrative Plant Biology, 59*(3), 158–163.

Zhang, L., Cheng, Z., Qin, R., Qiu, Y., Wang, J. L., Cui, X., Gu, L., Zhang, X., Guo, X., Wang, D., et al. (2012). Identification and characterization of an epi-allele of fie1 reveals a regulatory linkage between two epigenetic marks in rice. *Plant Cell, 24*, 4407–4421.

Zhang, X., Sun, J., Cao, X., & Song, X. (2015). Epigenetic mutation of RAV6 affects leaf angle and seed size in Rice. *Plant Physiology, 169*(3), 2118–2128.

Zheng, X., Chen, L., Xia, H., Wei, H., Lou, Q., Li, M., et al. (2017). Transgenerational epimutations induced by multi-generation drought imposition mediate rice plant's adaptation to drought condition. *Scientific Reports, 7*(1), 39843.

Glossary

Acclimation process by which an organism adjusts to a new environment or set of conditions. This adjustment can involve physiological and biochemical changes that allow the organism to survive and potentially thrive in the new environment.

Acquired traits phenotypic changes in an organism influenced by the environment.

Adaptation heritable trait shaped by evolutionary forces (primarily natural selection) that enhances reproductive success or survival in a given environment.

Alleles alternative versions of a gene or genetic *locus* that can exist at a specific chromosomal position within an genome.

Antisense transcripts non-coding RNA molecules transcribed from the opposite strand of a gene, exhibiting partial or complete overlap with the coding sequence.

Argonaute (AGO) highly conserved proteins that function as the catalytic core of the RNA-induced silencing complex (RISC), where they guide sequence-specific small RNAs to complementary RNA targets. Plants exclusively possess members of the AGO subfamily within the broader Argonaute family, which encompasses both AGO and PIWI subfamilies.

Cajal bodies dynamic nuclear suborganelles with a diameter of approximately 0.5 μm. These structures play a crucial role in the biogenesis of small ribonucleoproteins (snoRNPs) and contribute to the modification of non-coding RNAs (ncRNAs).

Chromatin complex of DNA and proteins (primarily histones) in eukaryotic cells. It serves primarily to package the genomic DNA into a compact structure suitable for the nucleus.

Chromatin-remodeling complexes (CRCs) ATP-dependent multiprotein assemblies capable of modulating gene expression. CRCs achieve this by dynamically altering chromatin structure through histone-DNA interactions, ultimately facilitating access of regulatory proteins to nucleosomal DNA.

Circular RNAs (circRNAs) class of non-coding RNAs (lncRNAs) that form covalently closed loops. Unlike linear RNAs, circRNAs lack a 5′ cap and 3′ poly(A)

© The Author(s), under exclusive license to Springer Nature Switzerland AG 2024
L. M. Vaschetto, *Epigenetics in Crop Improvement*,
https://doi.org/10.1007/978-3-031-73176-1_10

tail. They can function as competing endogenous RNAs (ceRNAs) and inhibit the interaction between miRNAs and their cognate target genes.

Climate change long-term alterations in temperature and typical weather patterns. These variations may potentially be attributed to natural phenomena (e.g., volcano eruptions) or anthropogenic activities, primarily the burning of fossil fuels (coal, oil, and gas) which releases greenhouse gases that trap heat in the atmosphere.

CRISPR/Cas9 gene editing technology for precise genome modifications. It takes advantage of a two-component system: the Cas9 nuclease enzyme and a guide RNA molecule. The guide RNA directs the Cas9 enzyme to a specific DNA sequence, where it introduces a targeted double-strand break. The cell's natural DNA repair mechanisms then mediate the desired genetic modification.

Demethylases enzymes that remove methyl (CH3) groups from their substrates (e.g., DNA and histones).

Dicer-like (DCL) conserved family of enzymes essential for biogenesis of noncoding RNAs (ncRNAs), which have been extensively characterized in several plant species, especially in the model plant *Arabidopsis*.

DNA methylation biological process that involves the addition of one methyl group (CH_3) to nucleotides in the DNA molecule (primarily cytosine), influencing gene activity without altering the underlying sequence.

DNA transposons also known as Class II transposable elements (TEs). Mobile DNA segments that utilize a "cut-and-paste" mechanism for transposition. This process involves the excision of the transposon from its original genomic location and subsequent integration at a new genomic site.

Ecological resilience capacity of an organism, population, or ecosystem to absorb disturbance and reorganize while retaining its essential functions and structures. This allows them to recover and return to a pre-disturbance state or transition to a new stable state.

Endonucleases enzymes that cleave internal phosphodiester bonds within nucleic acid strands.

Endoribonucleases specific class of endonucleases specialized in cleaving RNA molecules, either single-stranded or double-stranded depending on the enzyme's activity.

Enhancer sequences *cis*-regulatory elements that function by increasing the rate of transcription of nearby genes. These DNA sequences act as binding sites for transcription factors, which upon interaction, stimulate the recruitment of the transcriptional machinery.

Environmental challenge from a adaptative perspective, environmental challenge refers to the external pressures that organisms must overcome to ensure their survival and reproduction. In crop production, climate change presents a significant environmental challenge as it alters the conditions necessary for optimal growth and yield.

Environmental stresses diverse range of external factors that challenge the survival and fitness of populations. These stresses can be broadly categorized into two main types: abiotic (e.g., dehydration) and biotic (e.g., pests).

Epialleles variations in the epigenetic landscape at a specific gene *locus*. These variations can lead to heritable phenotypic differences persisting across multiple generations, even in the absence of changes in the underlying DNA sequence.

Epigenetic information systems Cellular regulatory mechanisms that integrate DNA methylation, diverse histone modifications (e.g., acetylation, methylation), and non-coding RNA pathways. These systems act cooperatively to establish and maintain heritable patterns of gene expression without altering the DNA sequence itself.

Epigenetic memory heritable modifications on chromatin, independent of the DNA sequence itself. These modifications can influence gene expression patterns and contribute to the long-term cellular properties and behaviors passed on to daughter cells.

Epigenetic reprogramming dynamic alterations of the epigenome, including DNA methylation and histone modifications, that occur during development to establish distinct cellular states.

Epigenetic switches self-reinforcing regulatory elements established through chromatin modifications. They can undergo spontaneous transitions between distinct heritable states, influencing gene expression patterns without altering the DNA sequence.

Epigenetic variation heritable modifications to the epigenome, independent of the underlying DNA sequence. These modifications can arise from various factors, including the environment, and influence gene expression patterns in cells, potentially impacting phenotypes across generations.

Epigenetics discipline that investigates how cells regulate gene activity beyond the DNA sequence itself. This field explores mechanisms, often mediated by chemical modifications, that influence gene expression without altering the underlying code.

Epigenome collection of chemical modifications to DNA, associated proteins, primarily histones, and non-coding RNAs (ncRNAs), that influence gene expression without altering the underlying DNA sequence. Epigenetic modifications can be influenced by various factors, including the environment, and contribute to the establishment and maintenance of cellular phenotypes. Some epigenetic modifications can be transmitted across generations through transgenerational epigenetic inheritance.

Epigenomics investigates the epigenome, the entirety of heritable chemical modifications on a cell's DNA that regulate gene expression without altering the DNA sequence itself.

Epimutations heritable alterations in gene expression patterns that occur independent of changes in the underlying DNA sequence. These modifications can have a profound impact on cellular phenotypes.

Epitranscriptomics also known as RNA epigenetics, refers to the study of post-transcriptional modifications (PTMs) of RNA molecules. These modifications, catalyzed by specific enzymes, introduce chemical alterations to RNA bases, impacting RNA structure, stability, and function.

Euchromatin open and decondensed regions of chromosomes enriched in genes. These regions typically harbor epigenetic modifications associated with active gene expression.

Evolutionary adaptation heritable modifications in an organism's structure or behavior that enhance its fitness within a specific environment. These modifications improve reproductive success and contribute to the persistence of the organism or its lineage over generations.

Evolutionary fitness organism's or population's ability to survive and reproduce within a specific environment. Individuals with traits that enhance these capabilities are considered more fit and have a greater chance of passing their genes on to future generations.

Exaptation trait that evolves for one function but is subsequently co-opted for a novel and advantageous purpose. Transposable element (TE) exaptation refers to the process where a newly inserted TE sequence (neo-insertion) confers a selective benefit to the host organism, even though the TE itself did not evolve for that specific function.

Gibberellins (GAs) class of plant hormones that orchestrate diverse developmental processes, including cell elongation and differentiation. They achieve this by stimulating the synthesis of RNA and proteins.

Gene expression cellular process by which the encoded information within a gene is translated into a functional product, primarily achieved through the transcription of RNA molecules. These RNAs can either serve as protein-coding templates or function as non-coding RNAs with diverse regulatory roles.

Genome stability faithful preservation of an organism's genetic information encoded within its DNA across generations. This fidelity is essential for the proper development, function, and inheritance of traits.

Genome all the genetic material present in one cell of a given organism.

Genomic imprinting phenomenon of epigenetic inheritance characterized by parent-of-origin-specific gene expression. In imprinting, the regulation of a gene or chromosomal region is determined by the sex of the transmitting parent, leading to monoallelic expression (expression from only one inherited allele).

Germination transition of a seed into a seedling, marking the resumption of plant growth. This process is contingent on favorable environmental conditions that break seed dormancy and initiate the developmental program leading to seedling establishment.

Green Revolution period of agricultural innovation from the 1960s that significantly increased food grain production, particularly in wheat and rice. This transformation was largely driven by the introduction of high-yielding crop varieties into developing countries.

H3 lysine 4 tri-methylation (H3K4me3) epigenetic mark, characteristically, but not invariably, linked to transcriptional activation. It has been associated with the transcriptional memory, transcription elongation, and stress responses.

H3 lysine 9 dimethylation (H3K9me2) epigenetic mark primarily associated with transcriptional repression of Transposable Elements (TEs) and repetitive sequences. This histone modification has also been associated to euchromatin and transcribed genes.

H3 lysine 27 trimethylation (H3K27me3) histone modification associated with transcriptional repression. It is deposited by Polycomb Repressive Complex 2 (PRC2) and contributes to the formation of condensed heterochromatin, leading to gene silencing. In plants, H3K27me3 is linked to developmental phase transitions (e.g., seed germination, and floral initiation).

Heterochromatin condensed genomic regions characterized by a predominance of repetitive DNA sequences. These regions exhibit limited accessibility to DNA-modifying enzymes, contributing to their transcriptionally repressed state. Compared to euchromatin, heterochromatin is generally more condensed and enriched with specific epigenetic marks, including H3 lysine 9 di/trimethylation (H3K9me2/3) and DNA (cytosine) methylation.

Hypomorphic allele Partial loss-of-function mutation. This reduction in gene activity can manifest at the level of protein or RNA expression, or in the protein's functional capacity.

Histone acetylases enzymes that catalyze the addition of acetyl groups ($COCH_3$) to histone proteins.

Histone deacetylases enzymes that catalyze the removal of acetyl groups ($COCH_3$) to histone proteins.

Histone modifications diverse array of epigenetic marks that influence chromatin structure and regulate gene expression through transcriptional activation or repression. They encompass methylation, acetylation, phosphorylation, ubiquitylation, and SUMOylation.

Histone-modifying enzymes enzymes that act on histone substrates after protein translation, catalyzing a diverse array of post-translational modifications. Cellular processes, including gene expression, are influenced by histone-modifying enzymes.

Histones family of positively charged proteins that tightly associate with DNA in the nucleus, facilitating its condensation into chromatin.

Independent traits genetic characteristics determined by different genes that segregate independently during meiosis. The independent assortment ensures that these traits are inherited from parents without bias towards co-inheritance.

Lamarckian evolution also called Lamarckism or Lamarckian inheritance, proposes that organisms can pass on traits acquired during their lifetime to their offspring. This concept contrasts with Darwinian evolution, where heritable traits arise through random mutations and are selected for through differential reproduction.

Linkage group collection of genes physically located on the same chromosome that tend to be inherited together during meiosis due to their close proximity.

Loci genomic regions physically separated, which can be associated with different genes or alleles.

Locus genetic region associated with a given gene.

Long non-coding RNAs (lncRNAs) diverse class of transcripts exceeding 200 nucleotides in length and lacking protein-coding potential. These molecules function as important regulators of gene expression through various mechanisms, including chromatin remodeling and epigenetic modifications. LncRNAs can act as either activators or repressors of gene transcription by interacting with DNA, RNA, and protein complexes. LncRNAs influence chromatin landscapes by recruiting enzymatic machinery that modifies nucleosomes, DNA methylation patterns, and histone post-translational modifications.

Matrix attachment regions (MARs) dinucleotide (AT)-enriched DNA elements that interact with the nuclear matrix, a proteinaceous network within the nucleus, shaping genome architecture. MARs are often found in gene-dense regions and are associated with transcriptional units.

Meiosis specialized cell division that reduces chromosome number in the parent cell, resulting in the formation of gametes. This reduction is essential for sexual reproduction to ensure the maintenance of ploidy levels in the offspring.

Metastable epialleles alleles with variable expression patterns in genetically identical individuals. This variability arises from epigenetic modifications established during early development and is thought to be particularly sensitive to environmental cues.

Methyltransferases enzymes that catalyze the process of methylation by adding methyl groups (CH3) to their substrates.

MicroRNAs (miRNAs) short, non-coding RNAs, typically around 22 nucleotides long, that function as regulators of gene expression, primarily by targeting messenger RNA (mRNA) with complementary nucleotide sequences. Although miRNAs mainly act at the post-transcriptional level to suppress protein translation, they can also control gene activity at the transcriptional level as well.

Miniature inverted-repeat transposable elements (MITEs) non-autonomous DNA elements found abundantly in plant genomes. These short sequences, typically ranging from 50 to 500 base pairs (bp), are characterized by terminal inverted repeats (TIRs) of 10–15 bp in length flanking the element. MITEs also possess short duplications (target site duplications or TSDs) at each end, marking the genomic insertion site.

Mitosis cell division process that ensures the faithful segregation of replicated chromosomes into two daughter nuclei. This equational cell division results in genetically identical daughter cells, maintaining the overall chromosome number. In higher organisms, mitosis gives rise to somatic cells.

Mutations alterations in the nucleotide DNA sequence. Mutations can have varying effects on an organism, ranging from beneficial adaptations to detrimental consequences or remaining neutral.

Natural selection major evolutionary force associated with differential survival and reproduction of individuals based on phenotypic variation. Organisms better adapted to their environments exhibit higher reproductive success, leading to the preferential transmission of genes associated with those advantageous traits. Over successive generations, this process drives directional evolutionary change within populations, resulting in divergence and the emergence of new species.

Non-coding RNAs (ncRNAs) diverse class of transcripts that lack protein-coding potential. ncRNAs are critical regulators across various biological processes in higher organisms, including plants and animals.

Nucleoli (singular: nucleolus) specialized, non-membrane-bound organelles within the nucleus of eukaryotic cells. These structures are essential for ribosome biogenesis, encompassing the synthesis, processing, and assembly of ribosomal subunits.

Nucleosome eviction cellular process during development that facilitates the opening of the chromatin structure. This increased accessibility of DNA sequences allows for the binding of essential regulatory factors, ultimately promoting gene expression programs, critical for differentiation.

Nucleosome occupancy density or enrichment of nucleosomes at specific genomic locations. This parameter reflects the presence or absence of nucleosomes over DNA sequences, such as those critical for transcription factor binding.

Nucleosome positioning precise location of individual nucleosomes along a DNA sequence.

Nucleosome basic structural unit of chromatin in eukaryotic organisms. It is comprised of DNA wrapped around an octameric complex of core histones (two each of H2A, H2B, H3, and H4).

Orthologs genes separated by the evolutionary process of speciation.

Paramutagenic allele an allele that induces a heritable silencing effect on the homologous *locus* known as paramutable.

Paramutation epigenetic phenomenon that describes the *trans*-interaction between alleles at a single *locus*. In paramutation, a silenced paramutagenic allele can induce a heritable silencing effect on a previously active paramutable allele. The silenced paramutable allele can itself acquire paramutagenic properties in subsequent generations.

Phenotypic plasticity the ability to adapt phenotypes, including behavior, morphology, and physiology, in response to changing environmental conditions.

Phosphorylation addition of a phosphate group (PO_4^{3-}) to a specific molecule.

Polycomb group (PcG) evolutionarily conserved multiprotein complexes essential for maintaining epigenetic inheritance. These complexes function as transcriptional repressors, particularly of master developmental genes, ensuring proper embryonic development by maintaining the repressed state of these genes across successive cell divisions.

Post-translational modifications (PTMs) covalent alterations following protein synthesis. These modifications play a critical role in expanding the functional diversity of the proteome by regulating protein activity, subcellular localization, and interaction with other molecules, including nucleic acids.

Priming inducible state of heightened preparedness for future stress responses. This enhanced readiness can be triggered by exposure to different biotic stresses (e.g., pests, pathogens), or specific environmental conditions.

Resilience ability to absorb disturbance and recover. This capacity is fundamental in the context of climate change, enabling species to adapt to changing environmental conditions and mitigate the negative consequences of climate events.

Retrotransposons also known as Class I transposable elements (TEs), mobile genetic elements that propagate via a "copy-and-paste" mechanism. This process involves reverse transcription of their RNA intermediate, followed by integration of the resulting cDNA copy into new genomic locations.

Ribonucleoprotein complexes (RNPs) macromolecular complexes found within living cells. They consist of two key components: (1) ribonucleic acid (RNA), which can be messenger RNA (mRNA), ribosomal RNA (rRNA), transfer RNA (tRNA), or other types of RNA molecules, and (2) proteins that bind to the RNA molecule, shaping its structure and influencing its function.

RNA interference (RNAi) biological pathway where double-stranded RNA (dsRNA) triggers sequence-specific gene silencing. In this pathway, dsRNA molecules are recognized and processed by the cellular machinery to target and degrade complementary messenger RNA (mRNA), effectively silencing the expression of specific genes. This evolutionary conserved mechanism can be harnessed for genetic engineering applications.

RNA-directed DNA methylation (RdDM) plant-specific epigenetic pathway. This pathway utilizes non-coding RNA molecules to guide *de novo* methylation of specific DNA sequences.

RNA-induced silencing complex (RISC) conserved multiprotein assembly that plays a critical role in RNA interference (RNAi)-mediated gene silencing. This complex functions in diverse pathways, including the suppression of transposable elements and defense against viral infections.

Seed dormancy physiological state that prevents germination even under favorable environmental conditions. This period of developmental arrest allows seeds to resist premature germination and ensures seedling establishment coincides with optimal conditions for growth and survival.

Small interfering RNAs (siRNAs) also known as short interfering RNAs or silencing RNAs, are double-stranded RNA molecules typically 20-25 nucleotides in length. At posttranscriptional level, these sequences (siRNAs) function by targeting and degrading messenger RNA (mRNA) with complementary nucleotide sequences, thereby silencing the expression of specific genes.

Small non-coding RNAs (sncRNAs) diverse collection of short transcripts, including microRNAs (miRNAs) and small interfering RNAs (siRNAs), that have less than 200 nucleotides in length and function as regulators of gene expression at the transcriptional and translational levels.

Single seed descent method strategy to accelerate plant breeding by isolating individual plants, harvesting a single seed from each, and combining these seeds to form the next generation.

Tandem repeats short nucleotide motifs directly adjacent to each other, forming arrays ranging from a few to hundreds of copies within a single *locus*.

Thermomorphogenesis morphological changes that plants undergo in response to non-stressful increases in ambient temperature. This adaptive mechanism helps plants to face altered environmental conditions in order to optimize their growth and survival.

Transactivation domains (TADs) protein regions that directly interact with general transcription factors, thereby stimulating gene activation.

Transgenerational epigenetic inheritance transmission of epigenetic information, including modifications that regulate gene expression, across multiple generations without altering the underlying DNA sequence.

Transgenerational stress memory protective effects that persist beyond one generation following the initial stress exposure.

Transgenerational stress responses inheritance of stress adaptation across generations.

Transposable element (TE) amplification also called TE proliferation, it is a well-established driving force for genome size expansion and evolution in plants. TE amplification is considered a primary mechanism for increasing plant genome size and contributing to broader evolutionary patterns within this kingdom.

Transposable element (TE) domestication co-option of TE-derived sequences by the host genome for beneficial cellular functions. This concept stands in contrast to the originally proposed 'selfish' function of TEs, highlighting their potential to be exploited by the host for beneficial purposes.

Transposable elements (TEs) also known as "jumping genes", they are mobile genetic sequences capable of relocating within the genome.

Ubiquitin small protein involved in the process of protein degradation.

Vernalization physiological process in plants where exposure to prolonged cold temperatures, typically mimicking winter conditions, is required for the initiation of flowering.

Index